GUIDE

TO FOOD TRANSPORT

FRUIT AND VEGETABLES

GUIDE

TO FOOD TRANSPORT

FRUIT AND VEGETABLES

Mercantila Publishers

© 1989 by Mercantila Publishers as

Layout and artwork:
Sylvester-Hvid A/S, Copenhagen

Printed in Denmark by
Specialtrykkeriet Viborg A/S

ISBN 87 89010 98 1

Distribution by:
Mercantila Publishers as
3 Malmøgade
DK-2100 Copenhagen
Denmark
Tel.: +45 31 26 06 21
Telefax: +45 31 26 16 21

CONTENTS

CONTENTS

CONTENTS

CONTENTS

PREFACE

The GUIDE TO FOOD TRANSPORT provides easy-to-follow guidelines to help assure that costly food cargoes reach their intended destinations in the best condition possible. Previous literature on this subject was quite technical, aimed at experts who already possessed some background knowledge. This new book is written so that it can easily be understood and used by all.

Improper or careless handling of fruit and vegetables during transportation can lead to damaged cargoes and extensive losses. The inadvertent destruction is often due to incorrect temperature and humidity settings during transport. It can also occur when products which ought to be stored separately are mistakenly put together.

The GUIDE TO FOOD TRANSPORT provides suggestions for avoiding such situations and the pursuing cargo loss. As such, it will be invaluable to people in all branches of the fruit and vegetable transportation industry i.e. shipowners, carriers, shipping agents, and consignors.

The first two chapters of the guide consist of a general introduction plus definitions and explanations of fundamental principles of fruit and vegetable transportation. Important questions regarding storage and transport conditions are also discussed.

This part of the book was compiled from information supplied by the Danish Bioteknisk Institut, by K. Porsdal Poulsen of the Technical University of Denmark (DTH) and by Mike Cowley of Cowley Industrial Consultants Ltd., UK. This information was then supplemented with know-how of shippers and manufacturers with both background and extensive experience in refrigerated transportation and container transport.

The main part of the book describes **one hundred** of the most important products in the fruit and vegetable category. It gives

PREFACE

advice regarding storage, perishability, handling and treatment. Information about the amount of heat generated by different types of cargo and about the maintenance of ideal temperature conditions is also included.

This latter section of the book was prepared by the Bioteknisk Institut of Denmark. The institute has collected relevant statistics and facts about fruit and vegetable cargoes throughout the years, and thus today possesses outstanding know-how in the field.

We would like to express our sincere thanks to all contributors who have made possible the publishing of this book and to the Mads Clausens Foundation for its encouragement and support.

July 1989   Mercantila Publishers as

1
CHAPTER

EXPLANATION OF INFORMATION ON COMMODITIES

NAME

The names of commodities in this book are given in English, Latin, French, German and Spanish. The naming system used for Latin names is stated. The official EEC translation, if available, has been used for the other languages. The most commonly used trade name has been selected for the remaining commodities. Widely used synonyms are also given.

DESCRIPTION OF PRODUCT

A brief description is given of the commodity's origin and an outline of the plant's appearance and growth. The appearance, size and structure of the edible parts are described, together with any special treatment needed during transportation of the product.

Any special characteristics of the product are listed, e.g. its use, hazards in handling or consumption, or content of any special substances.

MAJOR PRODUCERS

Some important producers are mentioned under this heading. Wherever it has been possible, the United Nations Food and Agriculture Organisation (FAO) production figures have been used as a basis. They therefore tend to reflect the largest producers rather than the largest exporters.

Statistical information has not always been available. In certain cases, therefore, groups of countries, or even continents, have been mentioned as important producers based on their significance in international trade.

STANDARDS

The information contained in this book covers fruit and vegetables for fresh consumption. Although much of the information may be used for commodities destined for industrial processing, such commodities are not dealt with here.

In this section distinction is made between mandatory standards and recommendations. Where there are standards pertinent to international trade, they are described in detail.

MINIMUM REQUIREMENTS

Minimum requirements emphasize specific conditions that should always be fulfilled. The minimum requirements will usually be included in international standards.

KEEPING QUALITIES

The commodity's keeping qualities are depicted in a shelf life diagram if adequate data have been available. Otherwise it is shown in the form of selected examples.

EXPLANATION OF INFORMATION ON COMMODITIES

SHELF LIFE
A shelf life diagram shows how the commodity's storage life depends on the chosen storage temperature. Some commodities deteriorate rapidly (mushroom, head lettuce), others are less susceptible (onions, white cabbage). The shelf life diagram is an approximate guide that depicts good normal storage conditions for sound commodities. Bad storage conditions or extremely good ones (for example CA-storage) will create a different relationship between shelf life and temperature. Similarly, there will be certain variations due to variety and growing conditions.

IDEAL DEMANDS
The ideal temperature and relative humidity ensure the optimum shelf life of the commodity.

RECOMMENDED TEMPERATURE
The thermometer depicted for a particular fruit or vegetable show the temperature intervals that are recommended for storage.

SENSITIVITY
The figure shows, by means of symbols, how sensitive the commodity is to storage at a different temperature (T) and relative humidity (RH) than that given under Ideal Demands.

The same symbols are used to show whether the commodity produces, or is susceptible to, ethylene.

Certain commodities (bananas, cucumber, etc) are very sensitive to storage at low temperatures. The resulting damage is called chilling injury and should not be mistaken for frost damage. Chilling injuries may occur long before the product freezes. In bananas, for example, it occurs when the temperature drops below 12°C.

Explanations of symbols
- – Insensitive/no ethylene production
- ★ Not particularly sensitive/insignificant ethylene production
- ★★ Sensitive/average ethylene production
- ★★★ Very sensitive/high ethylene production

HEAT PRODUCTION
The figure shows the heat production of the commodities caused by respiration at a temperature relevant to the product in question. Heat production is shown as a band due to the variations in varieties. The figures do not apply to climacteric fruits during the ripening process.

Climacteric fruit are those that have a marked period of biochemical activity, including a pronounced increased respiration rate and heat generation, usually triggered by an autocatalytic production of ethylene, but sometimes induced in ripening rooms by direct injection of the gas. The climacteric period for most fruits marks the change from growth to senescence. It should occur after the transport period and so is not considered here.

SPECIFIC WEIGHT
The specific weight is the expected density in bulk including packaging.

These figures can only be used as a guide since there is a wide range of variations globally due to differences in varieties and trimming and a profusion of packagings and stowage patterns. The figures given reflect products, trimming and packaging found in Europe.

The gross weight is given for packaged commodities, i.e. including packaging and pallet.

CONTROLLED ATMOSPHERES

GENERAL

In normal atmospheric air – comprising approximately 21% oxygen, 0.03% carbon dioxide and the rest nitrogen – commodities have unlimited access to that amount of oxygen necessary for respiration and, similarly, an unlimited ability to give off carbon dioxide. One method utilized to slow down the rate of respiration to extend the storage life of the commodity is to alter the composition of the air in the surrounding atmosphere, either by reducing its content of oxygen or by increasing its content of nitrogen, or a combination of both. Thus, commodities are exposed to partial suffocation that leads to a reduction in respiration rate.

TERMINOLOGY

Modified Atmosphere (MA)

The commodity itself has created a modified atmosphere, e.g. inside packaging, or in a carbon dioxide store without control.

Controlled Atmosphere (CA)

The atmosphere is controlled so that its composition is as desired. It can be adjusted if necessary.

The following terms are used under CA:
LO (Low Oxygen): When oxygen level is down to approximately 2%
ULO (Ultra-Low Oxygen): When oxygen level is below 2%.

RISKS

The reason for altering the composition of the atmosphere is to slow down the rate of respiration of the commodity but it is important that these alterations do not exceed the pre-set limits for the commodities. Too low oxygen concentrations will result in suffocation of the commodities due to lack of oxygen; too high concentrations of carbon dioxide will also cause suffocation because the commodity will be unable to emit carbon dioxide. There is a great risk involved for the majority of commodities in being exposed to oxygen concentrations below 1% and carbon dioxide concentrations above 10% for any length of time. The recommended combination for many fruits is about 3% oxygen and approximately 3% carbon dioxide.

PRACTICAL USE OF CA-STORAGE

Intensive research into CA methods is ongoing and the particular choice for each product is still subject to change, but the present utilization of CA is as follows:

CONTROLLED ATMOSPHERES

Utilization	Commodity	Remarks
Extensive	Apple	Long storage life
	Pear	6-8 months
	White Cabbage	
Increasing	Brussels Sprout	Long storage life
	Chinese Cabbage	Few months
	Kiwi Fruit	
	Leek	
Small quantities	Citrus Fruits	No commercial advantages
	Grape	have yet been demonstra-
	Papaya	ted on a large scale
	Pineapple	
	Mango	Marine CA container ship-
		ments over 30 days have
		been achieved

CHILLING INJURIES

Chilling injury is physiological damage which results from exposure of fruit and vegetables to temperatures below a critical threshold for each species – but above freezing temperature. It is primarily fruits and vegetables from the tropical and subtropical zones which are susceptible to chilling injury.

The extent of damage depends on the temperature, duration of exposure and the sensitivity of the fruit or vegetable. Commodities are divided into low, moderate and high sensitivity. With low sensitivity products it can take several weeks before chilling injury occurs, whilst high sensitivity products can suffer chilling injury after only a few hours. Chilling injury is usually only visible after subsequent exposure to temperatures higher than the critical threshold value.

TYPICAL SYMPTOMS OF CHILLING INJURY

Surface lesions – pitting, tissue breakdown e.g. courgette, mango and aubergine

Internal discolouration (browning) e.g. avocado, pineapple and banana

Failure of fruits to ripen e.g. tomato, papaya and mango

Abnormally accelerated rate of senescence e.g. grapefruit, sweet potato and cucumber

Increased susceptibility to decay e.g. pineapple, cucumber and cherimoya

Development of off-flavour and taste e.g. tomato, mango and papaya.

Commodity	Critical Threshold Temperature°C	Susceptibility
Aubergine	8	High
Avocado (1, 2)	7-8	High
Banana	12-14	High
Cherimoya	12-14	High
Courgette (1)	7-10	Moderate
Cucumber	7-12	High
Grapefruit (1)	10-15	Low/moderate
Guava	8-10	Moderate
Lemon	9-10	Low/moderate
Lime	8-10	Low
Melon (1, 2)	5-10	Low/high
Mango (2)	7-14	High
Papaya	7-8	High
Passion fruit	7-10	Low
Pepper (2)	6-8	Low/moderate
Pineapple (2)	7-13	High
Sweet potato	12-15	High
Tomato (2)	7-14	Moderate/High

(1) Dependent on species and/or variety
(2) More sensitive unripe than ripe

RESPIRATION OF FRUITS AND VEGETABLES

GENERAL

During growth in the fields, fruits and vegetables are supplied with sugar from the leaves through photosynthesis and with water and minerals through the roots.

On harvesting connection to the sources of this supply is severed, but the plant continues to respire, grow and mature and eventually to senescene, albeit under different conditions.

The fruit/vegetable has to rely on its own internal resources to generate the energy required for metabolism. The most pronounced condition is that the respiration process continues drawing water from the plant, water that can no longer be replaced from the roots.

The speed with which maturation occurs is an expression of how fast the commodity will age or break down and depends mainly on 3 factors – the commodity type and age at harvest, the temperature and the atmosphere surrounding the produce.

THE TYPE OF COMMODITY

There is a vast difference in the rates of respiration of various fruits and vegetables.

Commodity	Rate of Respiration	Storage Life
Potato, onion, head cabbage, apple, citrus fruit	Low	Long
Lettuce, cauliflower, leek, strawberry, pear, peach	Moderate	Short - moderate
Brussels sprout, spinach, artichoke	High	Short
Asparagus, broccoli, mushroom	Very high	Very short

Commodities that enter a dormant state when harvested, e.g. onion and potato, have low respiration rates, but commodities that are harvested during an active phase of growth, e.g. broccoli and asparagus, have very high respiration rates.

THE TEMPERATURE

In common with all other chemical processes, the respiratory process and other metabolic processes in fruits and vegetables are dependent on temperature. Within the biologically active temperature interval from 0-40°C, each time the temperature increases by 10°C there is an approximate doubling of the respiration rate.

NB! Certain fruits and vegetables are susceptible to physiological damage when exposed to temperatures below 8-12°C (see section on Chilling Injury).

THE ATMOSPHERE SURROUNDING THE PRODUCE

Oxygen is consumed during the respiration process. By limiting access to oxygen, it is possible to slow down the respiration rate and, in doing so, prolong storage life, but the total removal of all oxygen will result in suffocation. By increasing carbon dioxide content around the commodity, the same effect is achieved as by displacing the oxygen content. Too high a concentration of carbon dioxide will also damage the commodity so the correct balance must be obtained. The section on Controlled Atmospheres contains more information on these conditions.

The presence of ethylene in the atmosphere around the commodity will normally increase the rate of respiration and this will result in reduced storage life (see section on Ethylene).

ETHYLENE

GENERAL
Ethylene is a colourless gas which has a sweet ether-like odour. Physiologically, it is a hormone that accelerates senescence and the ripening process.

All plants and parts of plants produce ethylene in large or small quantities. The following commodities have a significant ethylene production rate: apple, avocado, netted melon, papaya, passion fruit, peach, pear and plum. It should also be noted that ethylene may be produced by burning coal, oil, petrol and tobacco.

The following product groups are especially sensitive to ethylene:

Commodity Group	Effects of Ethylene
Leafy vegetables	Turn yellow, russet spotting on leaves and abscission of leaves
cucumbers	Turn yellow and become soft
Unripe fruits	Accelerates ripening
Flowers	Wilt and/or drop off.

The effect of ethylene depends not only on the commodity itself, but also on temperature, exposure time and ethylene concentration. Many commodities, if exposed over lengthy periods, are sensitive to ethylene concentrations as low as 0.1 ppm (parts per million).

Ethylene or ethylene-like substances are used commercially to accelerate ripening of some fruits such as banana.

MONITORING OF ETHYLENE
Ethylene is monitored with various types of sensors, e.g. probes, flame ionization, photoionization and infra red absorption (IR). The first two also measure other hydrocarbon compounds in the air. In all the above methods, a sample of air is drawn through the sensor during monitoring. The ethylene concentration in the sample is then readily quantified by the discolouration of the sensor in the tube, which is discarded after each recording.

The other sensors can be used for continuous monitoring and the concentration is converted to electronic readings.

Stationary and portable monitoring equipment is available. A typical monitoring range is from around 0.01 ppm to approximately 50 ppm.

REMOVAL OF ETHYLENE

Ethylene can be removed by absorption with activated carbon, potassium permanganate, catalytic oxidation, in an ozone scrubber, or by fresh air ventilation. The latter is the easiest method.

Ethylene can be oxidized in a dry filter of potassium permanganate. The filter can only absorb a certain amount of ethylene, thereafter it must be renewed. Efficiency is reduced under humid conditions.

In catalytic oxidation the air must be heated to approximately 200°C before it can be led through the reactor bed where the ethylene is oxidized. The heating and subsequent cooling of the air results in a decrease in efficiency of the overall plant.

In an ozone scrubber ultra violet lamps produce ozone which oxidizes the ethylene. Efficiency at low concentrations is better than with catalytic oxidation, but there is a risk of accidental ozone leakage.

EVAPORATION AND WATER LOSS

GENERAL

Water loss is one of the main causes of deterioration in fruits and vegetables. There is , however, a slight difference between the various products in the amount of weight loss they can withstand before limpness and loss of crispness becomes so excessive that the products have to be discarded. Leek, cauliflower and carrot, for example, can withstand approximately 7% weight loss, apple, strawberry, pepper and mushroom, about 6%, whereas products such as lettuce and broccoli can only withstand 4%.

The margin of tolerance is small, but, on the other hand, there is a vast difference as to how prone they are to wilting. The surface structure of the commodities is the crucial factor. Certain products have shapes and surfaces that protect them well against dehydration, e.g. melon and tomato, whereas products such as lettuce, parsley and mushroom are almost unprotected. Dehydration can be retarded by the appropriate packaging of the products.

Water loss can also be hindered by maintaining the correct temperature and humidity in the storage room. This is due to the fact that evaporation of water from fruits and vegetables is primarily a physiological phenomenon arising from the difference in vapour pressure between the commodity and its surrounding atmosphere.

Usually, the relative humidity (RH) of a warehouse or a container is used to indicate the rate of evaporation the commodities are subject to, but RH is really a bad indicator because warm air may contain more water than cool air with similar RH. For example, the difference in vapour pressure (VPD) in air with 90% RH is 0.6 g/kg air at 5°C, and 1.2 g/kg at 15°C, i.e. twice as high.

Additionally, the water will tend to evaporate due to the increased temperature.

Even if RH is the same, dehydration of a commodity will be increased in a storage room or a container with a high temperature than a low one.

The increased tendency of water to evaporate at high temperatures also means that rapid precooling of the product to the required storage temperature is important.

To avoid unnecessary water loss in fruits and vegetables, it is advisable to store them under climatic conditions closest to the ideal temperatures.

PACKAGING

GENERAL

There are two types of packaging for fruits and vegetables, outer packaging which is the unit in which the products are handled, and the inner packaging which is the commodity's sales packaging.

OUTER PACKAGING

First and foremost, outer packaging is used as a transport unit. The packaging should protect the commodities from rough handling and at the same time be able to withstand a stacking height of up to 2.5 m. The packaging should not collapse due to high humidity, which is occasionally the case during transport.

The outer packaging should also allow adequate air flow so that the commodities in the inner packaging will maintain the desired temperature.

Ventilation holes are usually placed in the sides of the boxes, but as air flow is coming from beneath it is more effective if holes are positioned in top and bottom in a way that keeps air passage free.

In the retail trade outer packaging will often be used for commodity display purposes. The most utilized forms of outer packaging are cartons with various types of coatings that can withstand high humidity, sacks and boxes made of wood, fibreboard, plastic or other similar, often local, materials.

INNER PACKAGING

Packing of products in a sales package has disadvantages as well as advantages.

Advantages	Disadvantages
Protects against rough handling and contamination	Costs
	Impedes chilling
	Prevents trimming
Prevents dehydration	Danger of suffocation
Retards aging	
Facilitates sales distribution	

Choice of packaging should be based primarily on the individual requirements of the product.

Many wholesalers and retailers use packaging for information, logo and suggestions for use.

There is a wide range of materials and packaging forms available. The most widespread packaging is plastic film, either as bags or as wrapping, in combination with trays. Other packaging includes nets and paper bags, often coated with plastic.

There are no international standards for packaging. However, the materials used must comply with applicable national hygiene and non-toxicity standards.

REFRIGERATION

GENERAL
During transportation fresh fruits and vegetables preserve their quality best if their optimum temperatures are maintained. This normally implies that the commodities have to be precooled to this temperature before being loaded into the transport unit as refrigeration systems only have the capacity to cope with minor chilling tasks and to maintain product temperature.

If the commodities are highly perishable and, for some reason, have too high a temperature at the time of loading, they should be chilled as quickly as possible to avoid deterioration. The significance of refrigeration to each commodity is depicted in the relevant Shelf Life diagram.

COOLING REQUIREMENT
As mentioned, refrigeration systems in the transport unit only have limited refrigeration capacity and even a moderate quick-chill requirement of, for example, 1°C per hour demands an installed chilling capacity that is 2-3 times the normal.

A rule of thumb is that the chilling requirement for maintaining the necessary temperature per ton product is 0.25-0.35 kW plus 1.0 kWh per °C chilling.

Example:
A refrigerated container with a 15 ton load:
Chilling demand – 0.2°C per hour
Necessary chilling effect =
15 (0.30 + (1.0 × 0.2)) = 7.5 kW

DIMENSION REQUIREMENTS AND DEMANDS
The rule of thumb on cooling requirements is useful for a quick evaluation of an existing refrigeration system. However, it is insufficient in itself for use in designing new transport systems or for precise calculation of results that can be expected of existing systems.

The calculation of detailed dimensions, etc. may be based on many conditions and demands, the most important of which are as follows:

Conditions:
Product type(s) for the determination of maximum heat production and eventual critical temperature limits

Size of load and start temperature

Packaging and stacking pattern

Maximum and minimum external air temperature.

Demands:

Eventual chilling (°C per specified **time**) and lowest acceptable supply air **tempe-rature**.

Air temperature and relative humidity in the transport unit together with informa-tion on temperature limits.

OPERATION OF REFRIGERATION SYSTEM

The operation of the system depends on its construction and control system. It is therefore important that the handling agent is familiar with the directions sup-plied by the manufacturer.

During operation of the system in cold stores, it is necessary to distinguish be-tween the temperature that is recorded by the system's room thermostat and the cold air temperature measured where it is blown into the room. (delivery air) air)

This problem can be illustrated when the room thermostat sensor shows +1°C while cold air is being delivered into the room at -10°C to maintain the desired +1°C room temperature. Obviously, pro-duce stored near the -10 air inlet will suffer freezing or chilling injury.

On systems with inadequate supply air temperature control the problem may be minimized by moving the room thermo-stat sensor closer to the supply air inlet. This will, however, result in too high a temperature farthest away from the supply air inlet and the refrigeration unit will run in very short intervals. It is therefore important to be able to control supply air temperature and ensure that it will never be too cold.

In modern containers and trucks the temperature in the chilled produce range is invariably controlled by the supply air probe so this problem should not occur. (The return air probe measuring the room temperature is only used for deep frozen cargoes such as ice-cream and frozen meat. (see Temperature Control in chapter 2).

REGULATIONS

STANDARDS

To facilitate international trade in fresh fruits and vegetables, a wide range of quality standards is available. The EEC has a system comprising 28 standards which are mandatory for trade within the EEC.

Other international standards are not mandatory, but they may be useful as recommendations, e.g. OECD's Standards (Compendium of Applicable Standards, 1988) consisting of 53 standards, ECE standards (UN/ECE Standards for Fresh Fruits and Vegetables) consisting of 38 standards, and Central American standards (Norma Centroamericana), ZCAITI (Guatemala), which consists mainly of standards for fruit.

On a national basis there are only a few countries that have mandatory standards. Most countries have either no regulations at all or standards that are not mandatory. This includes US Grade Standards which comprise 87 standards to be used only as recommendations.

QUARANTINE REGULATIONS

Some countries have special quarantine requirements for imported commodities so as to avoid the spreading of various kinds of diseases and pests. This applies, for example, to Japan and the USA.

The American requirements are specified in Plant Protection and Quarantine Treatment Manual, USDA, 1976, which lays down the conditions applied to handling of perishable foodstuffs as well as the forms of documentation to be employed for importation into the USA.

Japan has forbidden the import of a wide range of commodities from certain regions, though certain exemptions may be granted if the commodities have been treated in a specific way approved by the Japanese authorities.

COLD TREATMENT

Cold treatment of Florida grown oranges and grape fruit is practised for export to Japan. Fruit flies e.g. Caribflies are killed when fruit is shipped at temperatures from 0.6°C to 2.2°C at a number of treatment days from 14-24.

Fresh Fruit Temperature	No. of Treatment Days
0.6°C (33°F)	14 days
0.8°C (33.5°F)	16 days
1.1°C (34°F)	17 days
1.4°C (34.5°F)	19 days
1.7°C (35°F)	20 days
1.9°C (35.5°F)	22 days
2.2°C (36°F)	24 days

CONTAMINATION

Most countries have regulations for governing maximum allowable content of pesticide residues in food commodities, but there is no common international set of rules.

There are common rules for acceptable content of pesticide residues within the EEC which are issued as Council Directive 76/895/EEC, Appendix II.

For information concerning other regions of the world it is necessary to refer to the individual country's regulations.

There are no common rules established that govern the acceptable content of heavy metals in fruits and vegetables within the EEC or other international regulations. Many countries have their own rules that are often very stringent.

SURFACE TREATMENT

There are no international standards for surface treatment practices. However, most countries have restrictions on this matter and normally allow only a few specific types of treatment. Many countries allow only coating preparations which are in themselves natural ingredients, whereas other countries also permit the use of additives.

Japan, for example, permits the use of polyvinylacetate (according to Specifications and Standards of Foods, Additives, etc.). In the USA a wide range of additives that have been approved for this purpose by the FDA is permitted. The EEC has no common regulations in this field.

As many countries legislate on the use of additives from an aesthetic point of view, a check must be made on the existing forms of surface treatment permitted in each country.

IRRADIATION

The use of ionizing radiation to extend the storage life of fruits and vegetables is a topic of great interest world-wide, and some controversy. There is no international legislation in this field, but FAO/WHO have drawn up a proposal for international standards.

This technique is employed in many countries, but is often not applied in practice due to tagging regulations and public opinion.

OTHER REGULATIONS

It should be noted that besides all the topics dealt with in this section, there are many other special regulations throughout the world. The use of chlorine in the washing treatment of fruits and vegetables is recommended in certain countries, but forbidden in others.

In certain countries the use of gases, e.g. carbon monoxide, in the prolongation of storage life of fruits and vegetables is permitted. The application of colouring to fresh fruits and vegetables is allowed in some countries, e.g. Japan, but there are no international regulations or outlines on this matter. It is appropriate, therefore, to refer to the particular country's regulations in each case.

VENTILATION

GENERAL

Ventilation during transport can either be internal ventilation with cold air to maintain a pre-set product temperature, or fresh air ventilation. Fresh air is introduced to remove carbon dioxide which has been released by the products during respiration and which can lead to oxygen deficiency in the transport unit.

FRESH AIR REQUIREMENT

This depends on the carbon dioxide production of the commodities and on the air-tightness of the container. For guidance, an estimate of fresh air requirement is 1-2 m^3 per ton/hour for the majority of calculations and, as a basis for further detailed calculations, an estimate of carbon dioxide production is approximately 0.2 litre CO_2 per 1 Watt respiratory heat produced.

Example:

1 ton cauliflower at 5°C breathes approximately 87 Watts/hour.

Produced carbon dioxide =
$0.2 \times 87 = 17.4$ litres per hour/ton. Necessary fresh air ventilation for the stabilization of carbon dioxide content in the container at maximum 1% =
$17.4 / (1000 / 100 \times 1) =$
1.74 m^3 per hour.

An effective and simple fresh air ventilation method is the installation of a small suction valve to introduce outside air to the suction side of the evaporator coil. The container should also be supplied with a similar valve for exhaust air.

Fresh air vents anywhere in the container will counteract too high a carbon dioxide content and oxygen deficiency. Effectiveness of the ventilation system depends on the operation of the refrigeration system and on the use of the container. The composition of the internal air should be checked periodically particularly with new systems. Most containers are fitted with a small pipe through the insulation, known as the CO_2 port, for sampling the inside air.

INTERNAL VENTILATION

Internal ventilation is a part of the cooling process for the removal of respiratory and transmitted heat and for the cooling of products.

Cold air is constantly circulated through the container to remove transmitted heat. During the cooling process cold air is brought into contact with the products, removing heat, water and gases.

Normally, there is one fan ducting that re-circulates the air through the evaporator coil and back to and around the container. In cases where product respiration rate is rather high and/or when warm products need to be cooled down quickly, air circulation through the products should be intensified. This can be done, for example, by using a special type of open packaging and stacking pattern. Under very demanding circumstances, there should be an air channel distribution system built into the stowage pattern of the produce to assist in the distribution of the air.

Air circulation around the product is a necessary evil and internal ventilation should therefore be reduced to the minimum that is necessary to maintain product temperature at the right level.

The temperature difference between cooling air and return air indicates whether the amount of air may be reduced or perhaps increased. If the temperature difference is under 1°C, consideration should be given to reducing the amount of air in the container, and if the difference, for example, is more than 3-4 °C, the amount of air should be increased.

Maximum air requirement should be, approximately, 600-700 m^3 per kW cooling effect and the amount should be capable of regulation within the range 1/3-1/1 amount of air. Infinitely variable ventilators are the recommended solution, but 2-3 step ventilators are also an acceptable solution.

MIXED LOADS

GENERAL
From a practical point of view, it is often necessary to stow several commodities in the same storage room or container even though they have different ideal demands of temperature and humidity. For short-term storage this is usually of no significance.

During long storage and transit periods serious problems may arise from mixed cargoes, not only because of incorrect temperature and RH, but also due to the emission of undesirable taste or odour-producing substances from the commodities, or on account of the emission of significant quantities of the undesirable ripening hormone, ethylene.

Given below are some directions for long-distance transport of mixed loads.

PRODUCTS WITH VARYING RELATIVE HUMIDITY DEMANDS
Transportation of mixed loads of commodities demanding varying relative humidities is inadvisable. The majority of commodities keep best at a high relative humidity, but onion, potato, yam and cape gooseberry can be adversely affected by it. On the other hand, the other commodities will be subjected to dehydration by evaporation if humidity is reduced.

UNDESIRABLE EMISSION OF ODOUR AND TASTE SUBSTANCES
Citrus fruit, potato, onion, leek, celery, garlic and other strong odour-producing commodities may give an off-flavour to more sensitive fruits and vegetables during mixed load transport. The most susceptible are apple, pear and melon.

ETHYLENE - PRODUCTION AND SENSITIVITY
Ethylene hastens the ripening of many commodities. Commodities that are particularly sensitive to ethylene should never, therefore, be transported together with commodities that emit significant quantities of ethylene.

Commodities that are very susceptible to ethylene are aubergine, banana, Brussels sprout, celery, cucumber, peach and pear.

commodities that are very susceptible to ethylene are aubergine, banana, Brussels sprout, celery, cucumber, peach and pear.

It should be noted that many commodities, e.g. banana, peach and pear, are very susceptible to ethylene and at the same time give off significant quantities of it. Ventilation of the storage room or container is therefore necessary. (For each product see diagram under Sensitivity).

CODE OF GOOD TRANSPORTING PRACTICE

GENERAL

It is important to bear in mind that quality is the factor that sells the final product. Quality can be maintained only if all links in the handling and distribution chain show equal consideration for the commodities. The transport chain therefore has the twofold responsibility of transporting the commodities and maintaining their quality during transport. We have summarised and called this task **quali-trans,** an acronym for the check list of actions or questions that have to be answered to ensure the successful movement of perishable produce.

RECEIPT

On receipt of a commodity intended for transportation the quality should be inspected. Whether the commodity is able to withstand the transit period is a factor that should be considered. (see chapter 2, Maturity and maturity indices).

Furthermore, the maturity and temperature of the commodity should be measured. High temperatures from the start can lead to disastrous outturn. Consideration of packaging material should be based on its ability to remove respiratory heat from commodities as well as its durability during transport.

TRANSPORT

A stacking pattern that ensures adequate circulation to remove respiratory heat should be selected. Stacking should depend on the amount of ventilation available.

Temperature recordings during transit depend on where the sensors are placed. Consideration needs to be given to the forms of documentation required and the adjustment of the probes accordingly. The aim should always be to give as correct a picture of the sequence of events as possible.

DELIVERY

The commodity should not leave the refrigeration chain during transport. Therefore, loading or reloading should only occur from refrigerator to refrigerator, with the temperature corresponding to that of the transport unit. Marked temperature fluctuations may result in moisture on the surface of the commodity and this may accelerate deterioration.

MONITORING EQUIPMENT

The recording equipment used should fit the task at hand and must be calibrated periodically. Inaccurate recordings provide no real information. Temperature probes should be calibrated in ice water and discrepancies should not exceed 0.5°C.

CODE OF GOOD TRANSPORTING PRACTICE

CHECK LIST

To facilitate the practical implementation
of quality-preserving transport, several
check points that distinguish **qualitrans**
are given below:

QUality	What is the quality like?	What was expected?
Allowed temperature	What is the temperature?	What was it supposed to be?
Laws and documents	What documentation is required?	
Incompatibility	Is it a mixed load?	Can the commodities be transported together?
Transport time	How long is the journey?	Is shelf life long enough?
Refrigeration capacity	What is the rate of respiration?	Is there adequate refrigeration capacity?
Air Exchange	What are the air exchange and ventilation requirements?	Is fan capacity sufficient?
Number of goods	How big is the load?	What quantity was expected?
Stowage	How will the load be stowed?	Is air circulation ensured?

2
CHAPTER

EDIBLE PLANTS – STORAGE AND TRANSPORT CONDITIONS

The majority of plant life is edible or can be used in one form or other as medicine, clothing or shelter. Humans normally eat only some 236 generic groups of plant. However, each generic type has numerous species and almost endless varieties, each of which is unique and requires a specific post harvest condition for optimum storage and transport. The banana alone has more than 1100 named clones, many of which are susceptible to post harvest handling in a slightly different way.

Handling the vast range of products is not as daunting as it may first appear, as 90% of all food eaten by man comes from 20 plant species of which the vast bulk are the staple foods i.e. cereal, yam and potatoes.

The number of different fruit and vegetable varieties in international trade is relatively small but growing rapidly. This book includes 100 products covering the major items and, by example, the majority of the remaining perishable foodstuffs in transit.

In a limited number of cases, we eat the whole plant, but in most cases we select one or two parts of the plant – such as
– sprouts, stems and leaves,
  inflorescences such as
– artichoke and cauliflower,
  partially developed fruit such as
– green bean, cucumber, okra,
  fully developed fruit such as
– apples and pears
  roots and tubers such as

– carrots, onions and potatoes,
  or even the seed such as
– dried beans.

Simply stated, both the growing environment and the part of the plant that is to be transported usually determine the type of treatment required in transit and the likely shelf life that can be expected.

In point of fact, the plant variety, the maturity at harvest and the part of plant to be transported all influence the ideal conditions. The combinations of these three important factors are almost endless, but fortunately for the operators, the storage and transport conditions that favour retarding the decay of plants after harvesting tend to fall into three distinct groups.

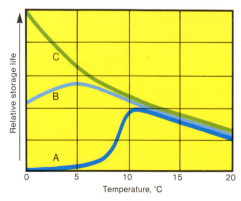

Figure 1 shows a simplified comparison of storage life for the three basic groups of produce.

EDIBLE PLANTS – STORAGE AND TRANSPORT CONDITIONS

Most plants from tropical and many fruits from subtropical zones are in group (A) and are susceptible to low temperature: leafy matter wilts, the structure of fruit is damaged. This is categorised as chilling Injury (see Chapter 1).

The second group (B) consists of products which are less susceptible to storage temperature. These tend to be temperate zone tubers, bulbs and large bulky fruit.

Products form group (C), such as the leafy part of plants from temperate zones, are best stored and transported at temperatures just above zero, with the best results obtained when the field heat is extracted from the plant as rapidly as is practicable.

TEMPERATURE CONTROL

In the majority of static stores and transport vehicles the temperature is maintained by a thermostat controlling the refrigeration machinery.

For local distribution and where there is a requirement for separate temperatures in the same vehicle, direct expansion of cryogenic gases can be used as the means of refrigeration. These systems usually use liquid nitrogen or carbon dioxide. The controls are particularly simple and reliable, but it is seldom cost effective. A thermostat injects the liquid gas behind a shield in the cargo space. The shield prevents produce coming in direct contact with the exceptionally low temperature of the gas.

The majority of transport refrigeration machines are relatively simple arrangements of a conventional circuit consisting of an expansion valve, evaporator coil, condenser coil and a compressor, with the thermal expansion valve providing the primary control to the circulating refrigerant.

To cool the cargo, the heat is extracted by the evaporator fans circulating air through the cargo and cooling the air by passing it over the refrigerant in the evaporator coil. The heat from the cargo space, plus the heat generated by the fan motors, is passed by the refrigerant to the condenser coil. Outside air is forced over the condenser by fans passing the heat from the coil into the atmosphere.

There is also a system for heating the air or to defrost the evaporator coil. Heat is introduced into the cargo either by electrical resistance heaters or by arranging the hot gas from the compressor to bypass the condenser and pass onto the cargo space by way of the evaporator coil.

Because there is a requirement to minimize the space occupied by the machine, the layout of pipes, wires and controls often gives a simple system the appearance of a complex machine.

The cutaway view **(Figure 2)** shows a typical layout of a refrigerated container. The internal air is circulated through the cargo and the machine. The conventional direction for the air to flow is shown by the arrows. This is known as bottom air delivery (Figure 3). In earlier containers and most long distance vehicles with little or no ducted floor, the air circulates in the reverse direction. This is known as top air delivery (Figure 4).

The fans (1) force the internal air through the evaporator coil (2) which cools it to the required temperature. The air then passes over the electric heaters (3) and over the delivery air thermostat (4) used by the controller and out into the container by way of the ducted floor (5). The most common form of ducted floor is known as T-grating, taking its name from the T-shaped cross section aluminium extrusions that form the floor.

The low pressure refrigerant gas passes heat from the evaporator (2) via the compressor (7) to the condenser coil (8) where heat is extracted by external air from the condenser fan (9). The now high pressure liquid refrigerant returns to the evaporator via the thermal expansion valve (not shown). On passing through the restriction in the valve from the high pressure to the low pressure side, the liquid expands to a gas and again absorbs heat from the air circulating cargo over the evaporator coil.

The temperature recorder (10) and/or the controller datalogger (11) measures and records the internal air temperature by a separate sensor (6), located in the return air passage.

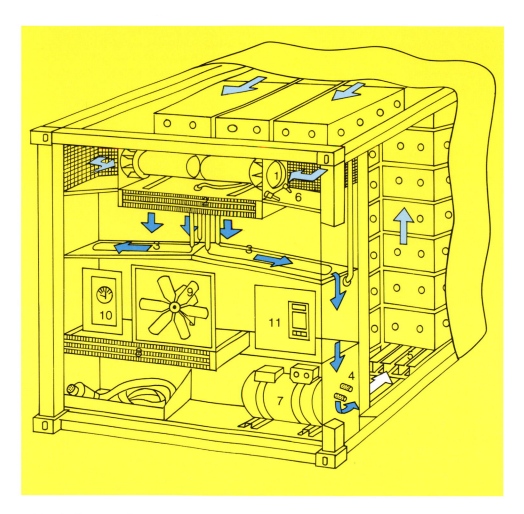

Fig. 2. *Cutaway view of refrigeration unit and container*

TEMPERATURE REQUIREMENTS

With perishable products of all types temperature is of prime importance, but the type of control is also important. The circulating air must not go below a pre-scribed limit otherwise the product will suffer freezing or chilling injury (see chapter 1), but the air temperature must also be as constant as is practicable. A fluctuating air temperature puts stress on the respiring plant, accelerating water loss and decay.

On simple machines used for relatively short journeys an on/off control device suffices, but on longer journeys, such as marine shipments, the compressor and fans are left running continuously and some form of capacity reduction is arranged to provide a constant supply air temperature within +/- a fraction of a °C.

CONTROLS

In addition to the primary control provided by the thermostatic expansion valve, there are a number of alternative additional controls providing a choice of sophistication and accuracy. With modern containers both a return and delivery air thermostat feed a control signal to an electronic, often computer based, controller. The controller adjusts the refrigerant circuit, the fans and the overall capacity of the machine to give a very precise air temperature at point of delivery. The controller often provides other

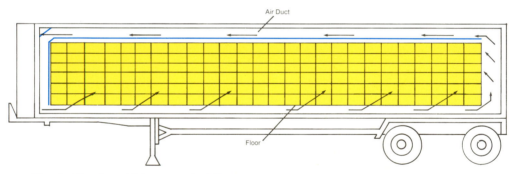

Fig. 3. *Trailer* with bottom air delivery

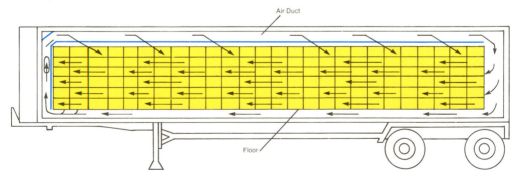

Fig. 4. *Trailer* with top air delivery

facilities, such as testing the machine's functions prior to loading and providing external signals for remote monitoring.

A common feature of all controllers is a programmed or selectable defrost feature. At the specified time interval, or in some cases when the pressure of the air increases due to the coil being blocked by ice, the controller will stop the fans and either by direct electrical heating or by diverting hot gas from the compressor will melt the ice. Little heat will enter the

cargo space but a considerable amount of water will drain away from the container. On a few of the latest models this de-icing condensate is collected, atomised and re-introduced into the cargo space. This helps to retain the humidity ratio.

HEAT SOURCES
Figure 5 illustrates the sources of the heat that has to be removed. It includes:
– heat conducted through the insulation
   from the outside air,

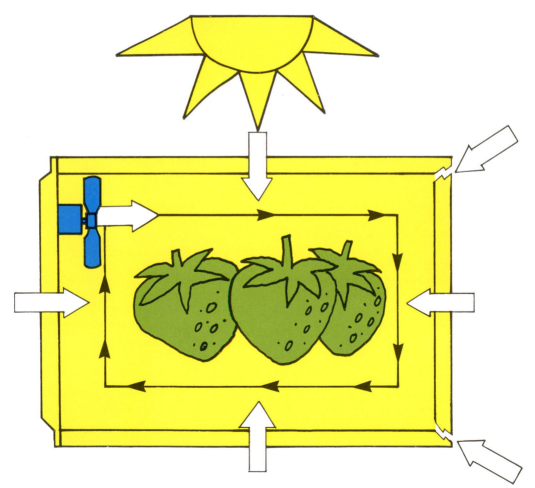

Fig. 5 *Heat Sources*

– heat absorbed from radiation from the sun or the road surface,
– heat from the respiring products,
– additional heat introduced by the ventilator and other air leaks,
– rejected heat from the evaporator fan and motor
– heat from any internal electric lights, if fitted.

It is the sum of this heat that must be removed by the cooling system and in such a way as to minimize the temperature difference of the air passing over the evaporator coil so as to avoid water condensing from the circulating air and the dry air subsequently dehumidifying the cargo.

PRE-COOLING

Transport equipment is designed to hold the product at a desired temperature. Container or truck units rarely have sufficient power to reduce the temperature of the mass of the cargo to the required level in a timely way. It is therefore necessary to pre-cool the cargo prior to loadings. (See Chapter 1 ''Cooling Requirements'' for capacity calculations).

Where pre-cooling facilities are not available and the product has to be loaded ''hot'', then every precaution should be taken to reduce the heat absorbed in the cargo prior to loading. This can best be achieved by harvesting in the night or early morning, keeping the product in the shade and in some circumstances by damping down with ice or water.

Loading cargo without precooling causes condensation problems. From **Table 1** and the section on the psychrometric chart it can be seen that as the temperature drops within the vehicle, water vapour will condense out of the air. A lot of this water will condense on the cold fins of the evaporator coil, forming ice which could restrict the air flow. Condensation will also collect on the cardboard cartons thus weakening the strength of the packaging, but the most damaging problem caused by condensate is that the free water provides a breeding ground for bacterial growth. The trend for packing vegetables in plastic overwraps or plastic bags means that, if the vegetable is not precooled, the subsequent cooling of the plastic results in free water inside the bag, providing a suitable environment for bacterial growth.

PACKAGING

There is a vast range of packaging from the simple bulk bin or net through to individual stylized polystyrene trays. The majority of packaging, though, is in the form of cardboard carton, tray or twopart box. Cheap fibreboard has a poor wet strength so there is a limit to the height to which cartons can be stowed in humid conditions

Tabel 1.

Influence of a temperature fluctuation ±1°C at various temperatures and humidity levels (the change in the relative humidity values is not symmetrical):

%RH	0°C	10°C	20°C	30°C
10 %RH	±0.8%	±0.7%	±0.6%	±0.6%
50 %RH	±3.7%	±3.5%	±3.2%	±3.0%
90 %RH	±6.6%	±6.3%	±5.7%	±5.4%

The design of the packaging should re-flect the fruit or vegetable, the required humidity level and the airflow type, the heat of respiration of the product and its susceptibility to bruising. However, pack-aging design type will often be deter-mined by the value of the product and be a compromise between protection, pro-motion and economy.

Nets and bags used for onions and pota-toes give good ventilation but poor stacking ability. Rigid boxes give good stacking but poor air circulation so ven-tilation holes are needed, provided they do not weaken the box strength. Plastic corners on board cartons markedly improve stacking ability and, at the same time, allow air to reach the product. Waterproof or water resistant cartons have to be used when items are packed with ice.

STOWAGE WITHIN THE TRANSPORT VEHICLE

It is self evident that whatever packaging is used, whether pallet, cardboard car-ton, net or bulk bin, they must be secure.

What is important is that while remaining secure, the packaging must allow air to circulate freely through the commodity, around the periphery of the container and in the area of the door. The import-ant criterion here is to have uniform di-stribution of air throughout the load. This requires the cargo to be uniformly stow-ed. Different sized packaging obviously dictates different stacking patterns as illustrated in Figures 6, 7 and 8. The di-

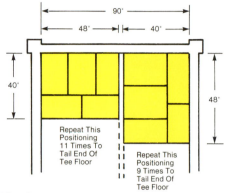

Repeat This Positioning 11 Times To Tail End Of Tee Floor

Repeat This Positioning 9 Times To Tail End Of Tee Floor

Fig. 6.

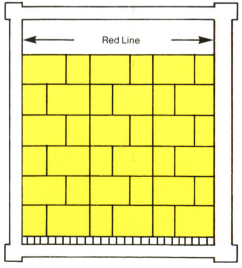

Fig. 7.

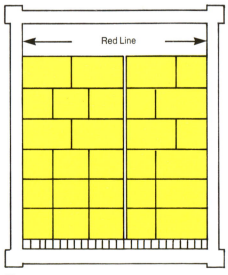

Fig. 8.

mensions of loose cartons will inevitably not be sized to fit the container exactly so the resulting gap must be spread out first to the left and the right.

With thin walls and insulation it is important that the cargo does not lean against the wall. Even with thick walled, well insulated vans or containers it is preferable that the cartons are held away from the wall by cargo battens built into the side walls.

The most common mistake is to load pallets or cartons up to ceiling height, restricting air flow along the return air passage over the top of the cargo to the evaporator fan. Marine containers are marked with a height limit. This should never be exceeded. (See Figure 7).

Where dissimilar sized packaging is used, or cargoes do not fill the container fully, it is recommended that additional empty cartons or some other material is used to fill up the void space so that the air passages remain uniform. In vehicles with canvas ducting on the roof, these should always be free and not restricted by the load.

Figure 9 shows the difference in cooling time for several arrangements of uniform size cartons.

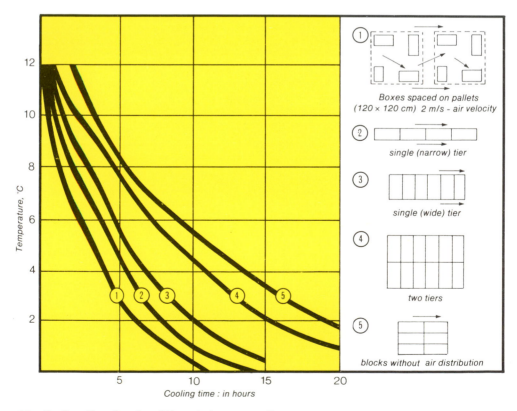

Fig. 9. *Cooling time for different stowage patterns*

The higher the resistance to the air pressure developed by the fans, the smaller the volume of air that will pass over the cargo and, subsequently, the lower the rate of heat exchanged between the air and the cargo. In an extreme case, the high resistance to air flow will mean that cargo will have relatively little or no air flowing over it. Conversely, in cargo stowed with large gaps and no resistance, the air will short circuit through the low resistance areas and return to the machine without cooling the bulk of the cargo. As stated earlier, the key to uniform cooling is uniform air distribution.

To achieve uniform air distribution within very long vehicles, different stowage patterns are sometimes used at the front and rear end to compensate for the pressure drop along the length of the vehicle. This is shown in **fig. 10**.

RELATIVE HUMIDITY AND THE PSYCHROMETRIC CHART

The fact that water is suspended in air is readily observed in our day to day life with mist, early morning dew and our

breath condensing in cold air. It is not important to understand the physics of water vapour other than that the relationship between temperature and the amount of water air can hold is non linear. Table 1 gives a few examples.

The psychrometric chart (Figure 11) is a way of displaying the conditions relating to water in air at a specific atmospheric pressure, usually sea level. The dry bulb temperature is shown on the horizontal line, the wet bulb temperatures are those on the lines falling to the right, the humidity percentages are the lines curved up and the vertical axis is the humidity ratio.

Relative humidity is a commonly used term to describe humidity of the air, but without knowing the corresponding wet or dry bulb temperature it has an imprecise meaning. It is more meaningful to talk of humidity ratio also known as absolute humidity that is the ratio of the weight of water in suspension in the air, to the weight of the dry air. This can best

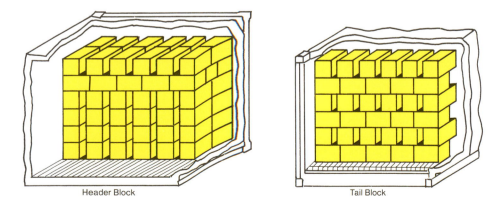

Header Block

Tail Block

Fig. 10. *Different stowage patterns within the container*

be explained by an example and by reference to the psychrometric chart **(Figure 11)**.

Relative humidity can be found by the point of intersection between the dry bulb and wet bulb temperature lines for example:

A dry bulb temperature of 25° C and a wet bulb temperature of 18° C will give a relative humidity of 50% (point A). By extending a line horizontally from point A to the right hand scale (point B), you will be able to read off the humidity ratio, 0.010 kg of water per kg of dry air.

If the air at this humidity ratio was to be cooled without changing its moisture content, it would form the horizontal line to the left, progressively increasing in relative humidity until it reached the 100% line at 14° C (point C). Any further

cooling would result in water condensing out of the air (dew would form). This is known as the dewpoint. To avoid free water forming on product, packaging or surfaces of the vehicle, the air circulating over cold surfaces must not be warmer or have a higher relative humidity than the corresponding point between C and A.

The important point to note here is that with perishable products requiring a high humidity at a temperature near 0° C, it is practically impossible to have the evaporator coil lower than circulating air temperature, say, 2° C without water condensing at the surface of the coil and subsequently lowering the relative humidity.

One way of keeping the humidity high is to ventilate the box with warm air from the outside. For example, air from out-

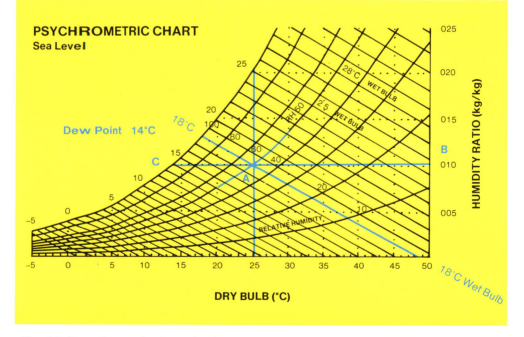

Fig. 11. *Psychrometric chart showing humidity and temperature relationship*

side with a dry bulb temperature of 25° C and wet bulb temperature of 18° C (point A) entering the vehicle or storage where internal temperature is 0° C and relative humidity of 95% will add moisture to the room as the air falls in temperature to 14° C (point C) when it will then loose water and will continue to do so until its temperature reaches the storage room temperature 0° C. The humidity ratio of this newly introduced air dropping from 0.010 to 0.004, will add 0.006 kg water for each kg of ventilated air added to the store. The limiting factors to using ventilated air to raise humidity is power consumption and having the external air in the appropriate condition.

MATURITY AND MATURITY INDICES

For the storage, transportation and distribution system to deliver fruit and vegetables at the required ripeness for the consumer, the harvesting of the living plant must take place at the correct point in the plant's natural development. The plant will continue to change after harvesting. It is common knowledge that the bananas eaten in North America and Europe are harvested in their green state and ripen en route or in storage at the point of destination. All other plants have similar but less pronounced characteristics.

As his measure of maturity, the banana plantation manager counts the number of days from the flower appearing to point of harvest. He then adjusts the day of harvest to take into account the journey time to final destination. For most other plants it is difficult to find such an easily quantifiable index of maturity, but they are available for most vegetables and fruits. Common indications are size, firmness, colour, and shape. Many of the regulations governing the sale of fruit and vegetables specify a maturity index of some sort. The problem is that objective maturity standards are rarely available as most rely on some form of subjective judgement.

Shippers, receivers and transport operators should develop a clear understanding of the maturity indices used for the products they carry. There is no technology available to overcome or reverse the process of ripening; only techniques to retard the process. If a cargo is already too mature when loaded for a particular journey to allow the goods to arrive at the required maturity, then a rejection, or claim for damages, by the receiver is inevitable.

SUMMARY OF GUIDELINES FOR DISTRIBUTION AND BULK TRANSPORT

DO
- always use refrigerated insulated vehicles, or containers where appropriate
- prechill the container before loading
- set the thermostate carefully and leave
  it to control the temperature
- check that the goods are within the
  correct temperature range before loading
- load and unload quickly to minimize
  exposure of chilled foods to ambient air
- segregate products to avoid crosscon-
  tamination
- allow sufficient air flow spaces between container surfaces and the product

dutc
- heat through coils during winter and
  other cold periods to avoid chilling injury and frost damage

DON'T
- expect refrigerated vehicles to chill
  products
- switch off the refrigeration plant at any
  time while products are in the vehicle
  except during loading or unloading
- transport unwrapped raw vegetables,
  fish or meat together in the same container, or in the same container as
  other chilled foods.

3

CHAPTER

APPLE

Latin:	Malus sylvestris Miller
French:	Pomme (commune)
German:	Apfel
Spanish:	Manzana

DESCRIPTION OF PRODUCT:

Apples originate most probably from Europe or the West Indies where they are still found growing wild.

Apples have been cultivated for several thousand years and now there are numerous varieties. They are cultivated in temperate regions of the world. The shape and colour of apples depend on the variety. Normally, apples are round with a 50-80 mm diameter. They are very sensitive to bruising which results in browning of the flesh.

MAJOR PRODUCERS:

The major producers of apples are U.S.S.R., China, U.S.A., France and West Germany.

STANDARDS:

For trade within the EEC apples must comply with EEC Standard No. 1. There are many recommendations, e.g. U.S. Grade Standards, but they are not mandatory in international trade.

MINIMUM REQUIREMENTS:

Apples should be intact, sound, clean and full-bodied. They should be free from visible foreign matter and should not emit any foreign smell or taste. Apples should be free from spots and signs of attack by pests or rot on the skin or in the flesh.

KEEPING QUALITIES:

Shelf life:

Storage depends on variety and method of storing.

Examples:

Golden Delicious	: 2°C	- 4 months
Golden Delicious	: CA-storage	- 8 months
Granny Smith	: 0°C	- 4 months
Granny Smith	: CA-storage	- 7 months
Jonathan	: 4°C	- 3 months
Jonathan	: CA-storage	- 5 months

Ideal demands:

0-6°C (32-43°F), 90-95% RH, depending on the variety

Recommended temperature:

°C	0	4	8	12	16
°F	32	39	46	54	61

Sensitivity:

T	RH	ETHYLENE PROD.	ETHYLENE SENS.	CHILLING INJURIES
★	★★	★★★	★★	★

Heat production:

HEAT PRODUCTION
W/ton

Specific heat: 3.64 kJ/kg x °C 0.87 Btu/lb x °F

Specific weight: Palletized boxes 320-380 kg/m³. Bulk 450-500 kg/m³

DESCRIPTION OF PRODUCT:

The apricot is a sub-tropical fruit and originates from Central Asia. It is a stone fruit approximately 3 cm in diameter and has a downy, pale orange surface. The stone is smooth.

Ripe apricots bruise very easily and highly perishable. They are therefore often picked unripe and then matured on the way to the consumer, losing some of their flavour in the process. After ripening they very quickly acquire a mealy consistency and taste.

MAJOR PRODUCERS:

The major producers of apricots are Turkey, Italy, U.S.S.R., Spain and France.

STANDARDS:

For trade within the EEC apricots must comply with EEC Standard No. 6a. Additionally, there are many recommendations, e.g. U.S. Grade Standards, but they are not mandatory in international trade.

MINIMUM REQUIREMENTS:

Apricots should be clean, sound and in-
tact. The fruits should bear no signs of decay or mould and they should be free of any foreign smell or taste. There should be no trace of mechanical damage. Apricots should be full-bodied, without bruises, surface defects or internal discolouration.

Latin:	Prunus armeniaca L.
French:	Apricot
German:	Aprikose
Spanish:	Albaricoque

KEEPING QUALITIES:

Shel life:

0°C, 90% RH, 1-2 weeks

20°C, 60%RH, 1-2 days

Ideal demands: 0°C (32°F), 90-95% RH

Recommended temperature:

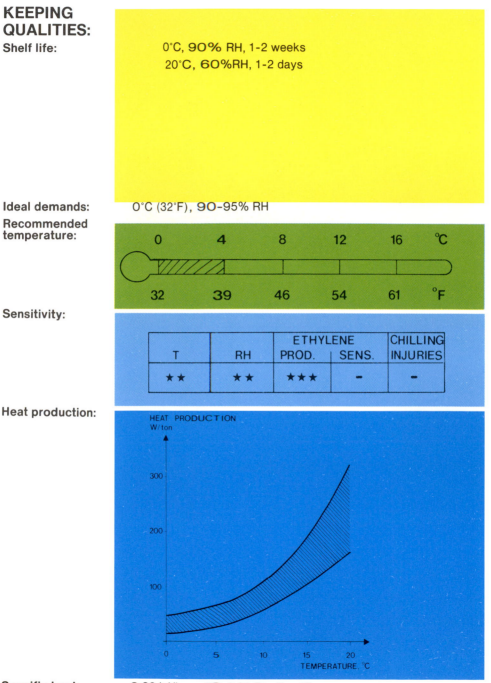

| 0 | 4 | 8 | 12 | 16 | °C |
| 32 | 39 | 46 | 54 | 61 | °F |

Sensitivity:

T	RH	ETHYLENE PROD.	SENS.	CHILLING INJURIES
★★	★★	★★★	–	–

Heat production:

HEAT PRODUCTION
W/ton

300

200

100

0   5   10   15   20

TEMPERATURE, °C

Specific heat: 3.68 kJ/kg x °C 0.88 Btu/lb x °F

Specific weight: Palletized boxes 300-350 kg/m³. Bulk 550-600 kg/m³

DESCRIPTION OF PRODUCT:

Globe artichokes came originally from the Mediterranean countries where they have been cultivated for over two thousands years.

It is a perennial, 2 metre tall thistle-like plant. Its unopened flower heads form the artichokes. They are harvested just before blossoming. The edible part is the base of the bracts and the floral receptacle.

Artichokes are about 10 cm in diameter and there are green, white and violet varieties.

MAJOR PRODUCERS:

The major producers of artichokes are Italy, Spain, France, Argentina and U.S.A.

STANDARDS:

For trade within the EEC artichokes must comply with EEC Standard No. 9. Additionally, there are many recommendations, e.g. U.S. Grade Standards, but they are not mandatory in international trade.

MINIMUM REQUIREMENTS:

Artichokes should be intact, fresh in appearance without any wilted bracts, sound, clean and not affected by rot, disease or pests. They should be free of foreign smell or taste. Artichokes should be firm and the bracts well closed. The colour must be typical of the variety.

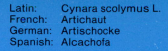

Latin: Cynara scolymus L.
French: Artichaut
German: Artischocke
Spanish: Alcachofa

Globe, **ARTICHOKE**

KEEPING QUALITIES:

Shelf life:

0°C, **95%** RH, 15-20 days
20°C, **60%** RH, 2-3 days.

Ideal demands: °C (32°F), 90-**95%** RH

Recommended temperature:

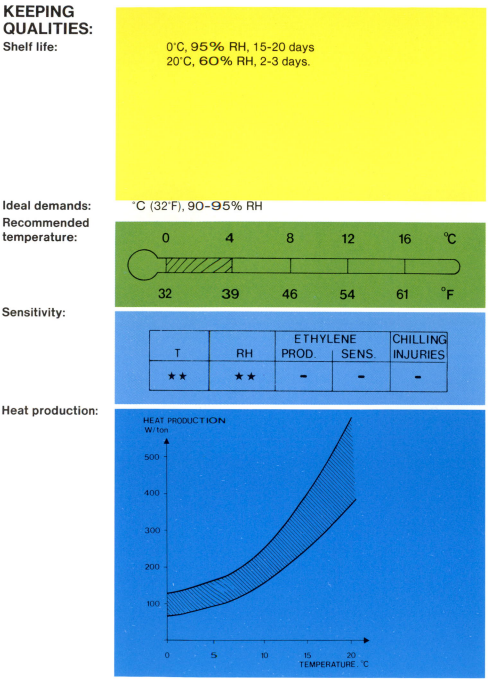

| 0 | 4 | 8 | 12 | 16 | °C |
| 32 | 39 | 46 | 54 | 61 | °F |

Sensitivity:

| | | ETHYLENE | | CHILLING |
T	RH	PROD.	SENS.	INJURIES
★★	★★	-	-	-

Heat production:

HEAT PRODUCTION
W/ton

500
400
300
200
100

0   5   10   15   20
TEMPERATURE. °C

Specific heat: 3.64 kJ/kg x °C 0.87 Btu/lb x °F
Specific weight: Palletized boxes 170-230 kg/m³

DESCRIPTION OF PRODUCT:

The asparagus is an ancient cultivated plant which has been well-known in the Mediterranean countries for many thousands of years.

White asparagus is a shoot which is cut off either before or just after the shoot breaks the surface of the soil. Should the shoot be cut later the tip of the shoot will discolour and this is considered a flaw. The most popular asparagus is long (up to 22 cm) and thick (approx. 2 cm in diameter).

MAJOR PRODUCERS:

The major producers of asparagus are Taiwan, France, Peru, Mexico and Spain.

STANDARDS:

For trade within the EEC asparagus must comply with EEC Standard No. 20. Additionally, there are many recommendations, e.g. U.S. Grade Standards, but they are not mandatory in international trade.

MINIMUM REQUIREMENTS:

Asparagus should be intact, fresh, sound, clean and full-bodied. There should be no trace of disease, rot or mould. The cut surface should be smooth. The shoots must neither be hollow, woody, frayed nor cracked.

KEEPING QUALITIES:
Shelf life:

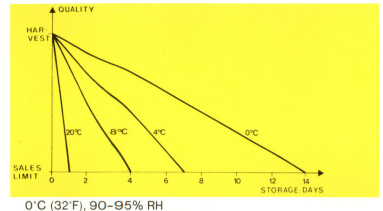

Ideal demands: 0°C (32°F), 90–95% RH

Recommended temperature:

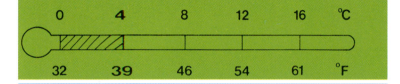

Sensitivity:

T	RH	ETHYLENE PROD.	SENS.	CHILLING INJURIES
★★	★★	–	★	–

Heat production:

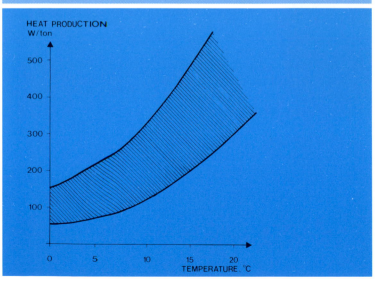

Specific heat: 3.93 KJ/kg × °C 0.94 Btu/lb × °F

Specific weight: Palletized cartons 270-350 kg/m³. Bulk approx. 600 kg/m³

DESCRIPTION OF PRODUCT:

Aubergines probably came originally from India or China. The plant is an annual, often thorny herb about 1.5 m tall. The fruits develop without pollination and can be harvested and consumed at any stage of growth before ripening. The pulp of ripe fruits is tough.

Aubergines come in many shapes, round and oval, and various colours such as white, red, yellow, purple and black. The pulp is spongy, pale and contains a number of light brown seeds.

MAJOR PRODUCERS:

The major producers of aubergines are China, Turkey, Indonesia, Italy and Tropical Africa.

STANDARDS:

For trade within the EEC aubergines must comply with EEC Standard No. 37. There are many other recommendations, e.g. U.S. Grade Standards, but they are not mandatory in international trade.

MINIMUM REQUIREMENTS:

Aubergines should be intact, sound, clean, firm and fresh. They should be adequately developed without the pulp being tough and with no significantly developed seeds. The fruits should be undamaged and free from rot, mould and any signs of disease. They should have no foreign smell or taste. Aubergines should be free from chilling damage which may result in sunken and discoloured parts on the surface of the fruit and in the pulp .

KEEPING QUALITIES:

Shelf life:

8-10°C, 90-95% RH, 10-14 days
20°C, 60% RH, 3-4 days

Ideal demands: 8-10°C (46-50°F), 90-95% RH

Recommended temperature:

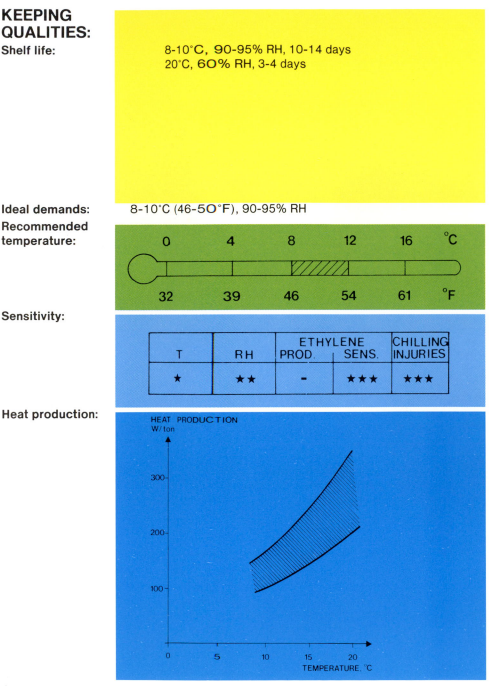

0	4	8	12	16	°C
32	39	46	54	61	°F

Sensitivity:

T	RH	ETHYLENE PROD.	SENS.	CHILLING INJURIES
★	★★	–	★★★	★★★

Heat production:

HEAT PRODUCTION
W/ton

300–

200–

100–

0   5   10   15   20   °C
TEMPERATURE, °C

Specific heat: 3.93 KJ/kg × °C 0.94 Btu/kg × °F
Specific weight: Palletized cartons 220-240 kg/m³. Bulk approx. 350 kg/m³

57

DESCRIPTION OF PRODUCT:

The avocado tree originates from South America. The fruit is pear-shaped and is normally green or brown. The size of the avocado varies from variety to variety and can be between 5-15 cm long. The skin is thin and usually wrinkled and leathery however some avocados have smooth skin. Its flesh is yellowish or light green and contains a large stone. The flesh of the avocado fruit discolours easily and it has a high fat content. Avocados have been used in the manufacture of soap.

A ripe fruit yields to slight pressure. The consistency of the ripe flesh is almost like that of butter.

MAJOR PRODUCERS:

The major producers of avocados are Mexico, U.S.A., Dominican Republic, Brazil and Indonesia.

STANDARDS:

Although there are no international standards for avocados, many recommendations can be found, e.g. U.S. Grade Standards, but they are not mandatory in international trade.

MINIMUM REQUIREMENTS:

Avocados should be clean, firm, intact and free from mechanical damage. They should be uniform in colour, without deformations, pitting or disease-infections. The colour should be typical of the variety. The flesh should have an even colour without brown spots. They should be free from foreign taste or smell. Chilling injury may result in discolouration of the flesh.

KEEPING QUALITIES:

Shelf life:

7°C, 90% RH,
2-4 weeks 20°C,
60% RH,
2-7 days.

Keeping quality depends largely on the fruit's stage of ripeness.
Avocados are susceptible to chilling injuries which result in brown discolouration of the flesh near the vascular bundle and around the stone. Depressions and failure to soften may also be due to chilling damage. Avocados from subtropical regions (Israel, South Africa) have 7°C as the lowest storage temperature and fruit from tropical regions 13°C.

Ideal demands: 7-13°C (45-55°F), 85-90% RH, dep. on var. and country of orig.

Recommended temperature:

0	4	8	12	16	°C
32	39	46	54	61	°F

Sensitivity:

	T	RH	ETHYLENE PROD.	SENS.	CHILLING INJURIES
	★★	★	★★★	★★	★★★

Heat production:

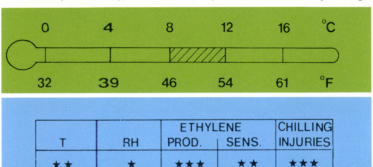

HEAT PRODUCTION
W/ton

TEMPERATURE, °C

Specific heat: 43.01 kJ/kg x °C 0.72 Btu/lb x °F

Specific weight: Palletized boxes 300-350 kg/m³. Bulk 500-550 kg/m³

BABY CORN

Latin: Zea mays L.
French: Maïs baby
German: Fingermais
Spanish: Maiz baby

DESCRIPTION OF PRODUCT:

Maize came originally from Latin America where it had been the staple crop of the Indians. It was the mature product that was mainly used as grain, but the use of the immature cobs was also known.

The maize plant has separate male and female flowers. The male flowers are at the top of the stem while the female flowers, that later become cobs, appear on the shoots of the stem. The »beard« at the top of the cob is the stigma. Each thread leads down to a maize kernel and, therefore, all the threads must receive pollen so as to ensure that all the kernels on the cob can develop fully. It is wind-pollinated.

Baby corn is the undeveloped cob whose centre has not yet become woody. The cobs are the size of a finger and are normally sold without their leaf wrappings.

MAJOR PRODUCERS:

The major producers of baby corn are U.S.A., China, Brazil, Thailand and Kenya.

STANDARDS:

There are no international standards for baby corn.

MINIMUM REQUIREMENTS:

Baby corn should be small and thin, the cobs whole and well developed without being woody. The cobs, which can either be with or without their leaf wrappings, should be well shaped and have distinctly formed maize kernels on the entire surface. The cobs should be firm and and full-bodied and the colour yellow or golden. They should be free from damage, spots or other visible defects, attack by disease or pests and from foreign smell or taste.

Latin: Zea mays L.
French: Maïs baby
German: Fingermais
Spanish: Maiz baby

KEEPING QUALITIES:

Shelf life:

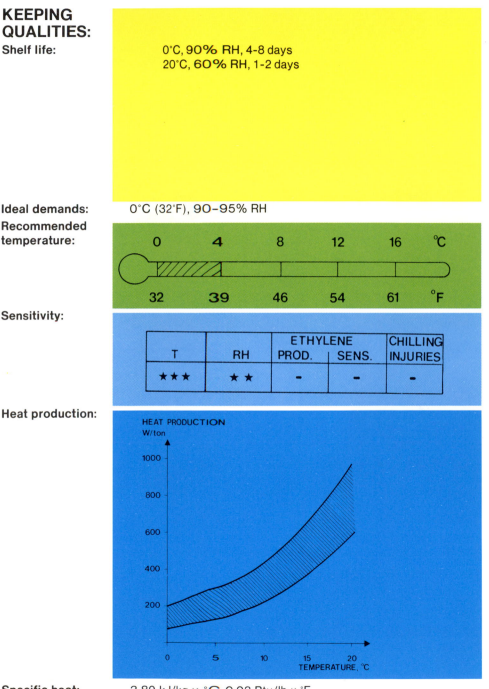

0°C, **90% RH**, 4-8 days
20°C, **60% RH**, 1-2 days

Ideal demands: 0°C (32°F), 90–95% RH

Recommended temperature:

	0	4	8	12	16	°C
	32	39	46	54	61	°F

Sensitivity:

T	RH	ETHYLENE PROD.	SENS.	CHILLING INJURIES
★★★	★★	–	–	–

Heat production:

HEAT PRODUCTION
W/ton

TEMPERATURE, °C

Specific heat: 3.89 kJ/kg x °C 0.93 Btu/lb x °F
Specific weight: Palletized cartons 200-250 kg/ m³

DESCRIPTION OF PRODUCT:

The banana is considered to be one of the world's oldest cultivated plants. It originates from tropical South East Asia where wild varieties can still be found. The banana plant is often called a palm, but it is actually an herbaceous crop and has such closely set leaves that they form a trunk. Within 6 months the plant grows to its full height, which is 4-10 m, and within the same year it blossoms and sets fruit. Harvesting is accomplished by cutting down the tree and picking the unripe bananas. When they arrive at their destination the bananas are artificially ripened depending on sales demand.

MAJOR PRODUCERS:

Major producers of bananas are Brazil, India, the Philippines and Ecuador.

STANDARDS:

There are no international standards for bananas.

MINIMUM REQUIREMENTS:

Bananas should be whole and sound without any deformities. The skin should not be spotted or bear any signs of mechanical damage. It should be intact and not expose the flesh. Overripe, rotten and mouldy bananas should be discarded. Bananas should have a smell and taste characteristic of the stage of maturity. They should be free from chilling injury which can result in a brown discolouration of the fibres.

Latin:	Musa X paradisiaca L.
French:	Banane
German:	Banane
Spanish:	Banana

BANANA

KEEPING QUALITIES:

Shelf life:

Green bananas: 13°C, 2-3 weeks.
A few varieties, e.g. Gros Michel, can be stored at
12°C for a shorter period. At 20°C the keeping quality is 4-8 days.

Yellow bananas: 13°C, 3-6 days.
At 20°C the keeping quality is 1-2 days.

Ideal demands: 12-14°C (54-57°F), 90-95% RH, depending on the variety.

Recommended temperature:

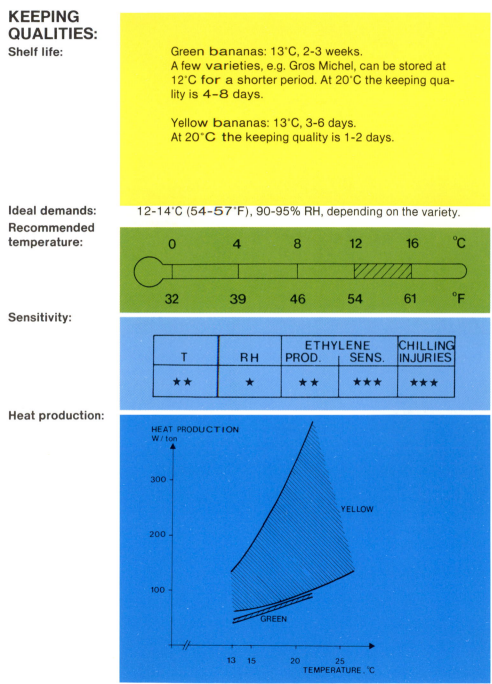

Sensitivity:

T	RH	ETHYLENE PROD.	SENS.	CHILLING INJURIES
★★	★	★★	★★★	★★★

Heat production:

Specific heat: 3.35 kJ/kg x °C 0.80 Btu/lb x °F
Specific weight: Palletized cartons 320-350 kg/m³

DESCRIPTION OF PRODUCT:

The bean is presumed to have originated in Central America where it has
been known for at least 6000 years.

The bean is usually a twining herb. The
edible pod is fleshy and either flat or
round with a considerable number of undeveloped seeds. As the seeds develop
the pods gradually become less fleshy.
There are many different varieties with
various shapes and sizes, the most
popular being the string-less bean, i.e.
the pod being free of fibres. Beans are
normally green, but yellow and violet varieties also exist.

MAJOR PRODUCERS:

The major producers of beans are China,
Turkey, Italy, Spain and Egypt.

STANDARDS:

For trade within the EEC beans must
comply with EEC Standard No. 16. Additionally, there are many recommendations, e.g. U.S. Grade Standards, but
they are not mandatory in international
trade.

MINIMUM REQUIREMENTS:

The pods of the beans should be intact,
clean and sound. They should not be affected by disease, pests, rot or mould
and should be free of foreign smell or
taste. The pods should be sufficiently
developed, but without such seed development that would make the beans
unfit for human consumption. The pods
must be full-bodied and have a fresh
appearance.

KEEPING QUALITIES:
Shelf life:

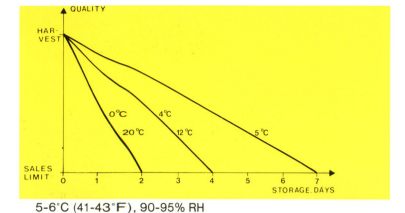

Ideal demands: 5-6°C (41-43°F), 90-95% RH

Recommended temperature:

	0	4	8	12	16	°C
	32	39	46	54	61	°F

Sensitivity:

T	RH	ETHYLENE PROD.	SENS.	CHILLING INJURIES
★★★	★★★	-	-	★★

Heat production:

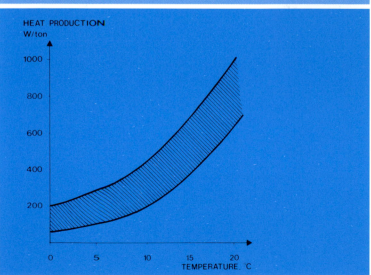

Specific heat: 3.85 kJ/kg x °C 0.92 Btu/lb x °F
Specific weight: Palletized cartons 240-280 kg/m³. Bulk approx. 400 kg/m³

65

(Bitter cucumber or Karella)

BITTER GOURD

Latin: Momordica charantia L.
French: Margose
German: Carella or Bittere Springgurke
Spanish: Cundeamor

DESCRIPTION OF PRODUCT:

Bitter gourd probably originates from Asia, but its exact provenance is unknown. After it was brought to Brazil by slave ships bitter gourd spread to all tropical regions.

The plant is an annual climber which can be up to 5 m long. The stem is 5-edged and very wrinkled. The gourd has an irregular shape and the surface is furrowed with wartlike protrusions. The flesh contains a large number of brown, rough seeds. The green, unripe fruit with pale yellow pulp is used as a vegetable. A ripe bitter gourd is yellow and the pulp red. There are various sized bitter gourds, ranging from approx. 8 cm to about 30 cm in length. As the name suggests, bitter gourd is very bitter.

MAJOR PRODUCERS:

Bitter gourd is produced in tropical regions throughout the world, with India, Indonesia, Malaysia and Thailand as the major producers.

STANDARDS:

There are no international standards for bitter gourds.

MINIMUM REQUIREMENTS:

Bitter gourd should be intact, clean and sound. There should be no signs of attack by disease or pests. The fruit should be firm, without any bruises arising from mechanical damage. It should be green or yellow depending on the desired stage of ripeness.

Latin: Momordica charantia L.
French: Margose
German: Carella or Bittere Springgurke
Spanish: Cundeamor

(Bitter cucumber or Karella)
BITTER GOURD

KEEPING QUALITIES:

Shelf life:

10°C, 90% RH, 2-3 weeks 20°C, 60% RH, 3-5 days

Ideal demands: 8-10°C (46-50°F), 90-95% RH

Recommended temperature:

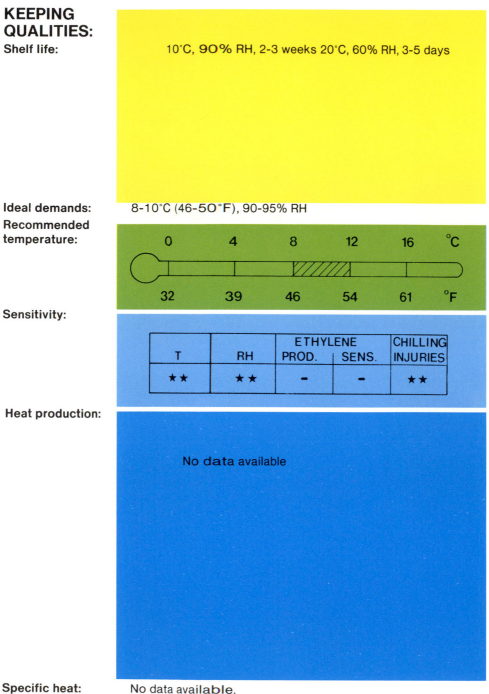

0	4	8	12	16	°C
32	39	46	54	61	°F

Sensitivity:

T	RH	ETHYLENE PROD.	SENS.	CHILLING INJURIES
★★	★★	–	–	★★

Heat production:

No data available

Specific heat: No data available.

Specific weight: Boxes, pallets 300-380 kg/m^3. Bulk approx. 550 kg/m^3

BLACK RADISH

Latin:	Raphanus sativus L. var. niger (Mill.) S. Kerner
French:	Radis noir
German:	Rettich
Spanish:	Rábano negro

DESCRIPTION OF PRODUCT:

Black radish is presumed to have originated from Asia Minor and spread to other countries, including China and Japan. Black radish is one of the most ancient vegetables. The slaves who built Cheops Pyramid around B.C. 2700 are said to have had black radish as an integral part of their diet.

Black radish should be intact, clean, full-bodied and look fresh. It should not be closely related to radish, but is much bigger, has a thicker skin and tastes a little more pungent. It comes in many colours such as white, red and black, and shapes ranging from round, conical to oblong and cylindrical. The most commonly known are the oblong, red and the round, black varieties. The strong taste arises from its content of mustard oils.

MAJOR PRODUCERS:

The major producers of black radish are China, Japan, West Germany, Italy, France and the Netherlands.

STANDARDS:

There are no international standards for black radish.

MINIMUM REQUIREMENTS:

Black radish should be intact, clean, full-bodied and look fresh. It should not be woody, hollow or have spongy tissue or other internal defects. It should be free from mechanical damage, signs of disease or attack by insects. The roots should be well developed with a colour typical of the variety. If black radish is sold with leaves, they should be fresh and green.

Latin: Raphanus sativus L. var. niger (Mill.) S. Kerner
French: Radis noir
German: Rettich
Spanish: Rábano negro

BLACK RADISH

KEEPING QUALITIES:
Shelf life:

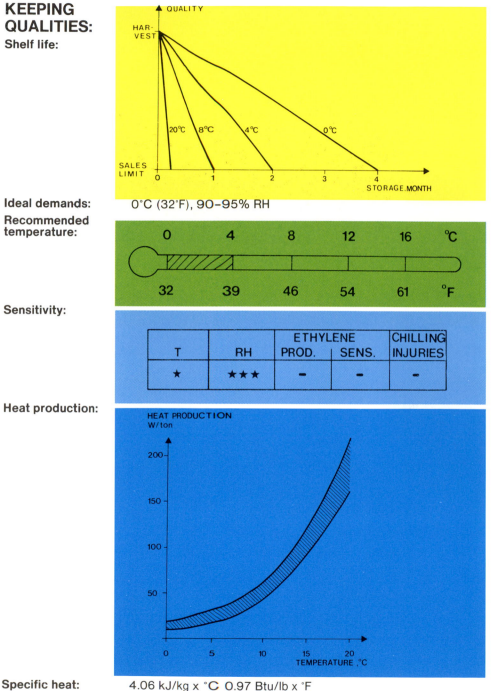

Ideal demands: 0°C (32°F), 90–95% RH

Recommended temperature:

Sensitivity:

T	RH	ETHYLENE PROD.	SENS.	CHILLING INJURIES
★	★★★	–	–	–

Heat production:

Specific heat: 4.06 kJ/kg x °C 0.97 Btu/lb x °F
Specific weight: Boxes, pallets 280-350 kg/m³. Bulk approx. 250 kg/m³

BLUEBERRY (Whortleberry)

Latin: Vaccinium corymbosum L.
French: Bleuet (Canada) or Myrtille
German: Heidelbeere or Blaubeere
Spanish: Arándano or Mirtillo

DESCRIPTION OF PRODUCT:

There are many varieties of blueberry.
The American variety is a tall bush (up to
1.5 m) which bears clusters of large
blueberries.

Wild bushes in Europe bear individual or
very small clusters of small berries.
These bushes are no taller than
30-40 cm.

The blueberry bush is a perennial which
bears fruit after about 4 years and can
have a productive life of some 75 years.

MAJOR PRODUCERS:

The major producers of blueberries are
U.S.A., the Netherlands, West Germany
and Poland.

STANDARDS:

There are no international standards
for blueberries. However, there are many
recommendations, e.g. ECE Standard
FFV-07 and U.S. Grade Standards, but
they are not mandatory in international
trade.

MINIMUM REQUIREMENTS:

Blueberries should be intact, fresh and
full-bodied. They should be evenly ripe
and include little leaf and stalk matter.
The berries should be free from mould,
rot and attack by insects. There should
be no signs of mechanical damage or
skin cracking. Blueberries should have a
characteristic taste and be free of any
foreign smell or taste.

70

Latin: Vaccinium corymbosum L.
French: Bleuet (Canada) or Myrtille
German: Heidelbeere or Blaubeere
Spanish: Arándano or Mirtillo

(Whortleberry) **BLUEBERRY**

KEEPING QUALITIES:

Shelf life:

0°C, 90% RH, 10-14 days   20°C, 60% RH, 1-2 days

Ideal demands: 0°C (32°F), 90–95% RH

Recommended temperature:

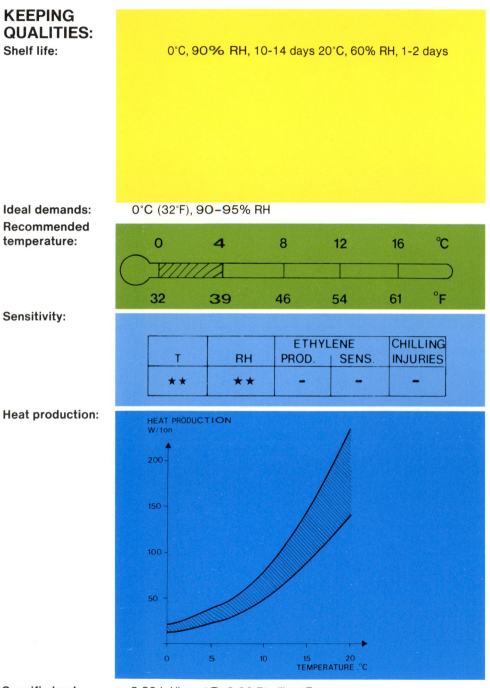

	T	RH	ETHYLENE PROD.	SENS.	CHILLING INJURIES
	★★	★★	-	-	-

Heat production:

Specific heat: 3.60 kJ/kg x °C   0.86 Btu/lb x °F

Specific weight: Palletized cartons 250-350 kg/m³. Bulk approx. 580 kg/ m³

BROCCOLI (Calabrese)

Latin:	Brassica oleracea L. convar. botrytis (L.)
French:	Chou brocoli or Calabrais
German:	Brokkoli or Spargelkohl
Spanish:	Broculi or Brécoles

DESCRIPTION OF PRODUCT:

Broccoli has been developed from wild cabbage that grow all over the Mediterranean region. It has been known for centuries in Southern Europe and, especially in Italy, grown on a large scale.

Broccoli resembles cauliflower in structure. The head, which is relatively open in structure, comprises an inflorescence of active flowers at budding stage. When the main spear is cut off side spears will appear. As these are not as big as the main spear, they are often sold in bunches.

Broccoli is much more sensitive than cauliflower to unfavourable humidity and temperature conditions.

MAJOR PRODUCERS:

The major producers of broccoli are U.S.A., U.K., Italy and Spain.

STANDARDS:

There are no international standards for broccoli. On the other hand, there are many recommendations, e.g. U.S. Grade Standards, but they are not mandatory in international trade.

MINIMUM REQUIREMENTS:

Broccoli should be sound, clean and free of any foreign smell or taste. I should be full-bodied, firm, fresh and green without budding yellow flowers. The attached leaves should be fresh and green. They should be free from defects in development, and attack by disease and mechanical damage, especially to the spear itself. It should also be free from larvae, plant lice and other pests. Broccoli should be cleanly cut. If a spear has been divided into florets, the cut surfaces should not be frayed. Spears with no florets are not acceptable.

Latin:	Brassica oleracea L. convar. botrytis (L.)
French:	Chou brocoli or Calabrais
German:	Brokkoli or Spargelkohl
Spanish:	Broculi or Brécoles

(Calabrese) **BROCCOLI**

KEEPING QUALITIES:

Shelf life:

0°C, 90% RH, 1-2 weeks 20°C, 60% RH, 1-2 days

Ideal demands: 0°C (32°F), 90–95% RH

Recommended temperature:

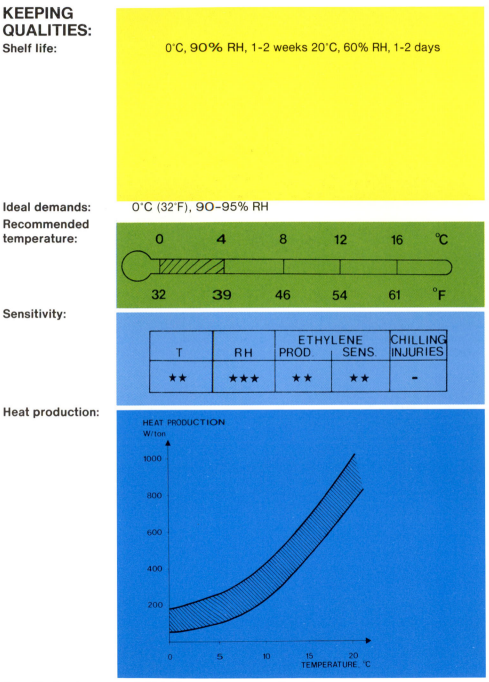

	T	RH	ETHYLENE PROD.	SENS.	CHILLING INJURIES
	★★	★★★	★★	★★	-

Sensitivity:

Heat production:

HEAT PRODUCTION
W/ton

TEMPERATURE, °C

Specific heat: 3.90 kJ/kg x °C 0.93 Btu/lb x °F

Specific weight: Palletized boxes 150-190 kg/m³. Bulk approx. 300 kg/m³

BRUSSELS SPROUT

Latin : Brassica oleracea L.
French: Chou de Bruxelles
German: Rosenkohl
Spanish: Col de Bruselas

DESCRIPTION OF PRODUCT:

Brussels sprouts are the youngest of the different types of cabbage. Their origin is unknown, but are believed to have been hybridized in Belgium around the year 1500. From there they spread to the rest of Europe, but were only used on a small scale. Large scale cultivation did not begin until the 19th century.

The plant is a biennial. During its first year it develops a 1 metre tall vertical stem. At each leaf the stem bears a lateral bud which stores the plant's nutrients. These buds, Brussels sprouts, can be about 3 cm in diameter when they are harvested, either manually or by mechanical cutting of the stems.

MAJOR PRODUCERS:

The major producers of Brussels sprouts are U.K., France, the Netherlands, Belgium and U.S.A.

STANDARDS:

For trade within the EEC Brussels sprouts must comply with EEC Standard No. 25. There are many recommendations, e.g. U.S. Grade Standards, but they are not mandatory in international trade.

MINIMUM REQUIREMENTS:

Brussels sprouts should be intact, fresh, sound, free from soil and foreign smell or taste. They should be free from signs of attack by insects, disease, rot, mould or from frost and mechanical damage. The colour should be green without yellow leaves. The small heads should be firm.

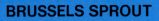
KEEPING QUALITIES:
Shelf life:

QUALITY

HAR-
VEST

20°C 8°C 4°C 2°C 0°C -1°C

SALES
LIMIT 0   5   10   15   20

STORAGE, DAYS

Ideal demands: -1-0°C (30-32°F), 90-95% RH

Recommended temperature:

0	4	8	12	16	°C
32	39	46	54	61	°F

Sensitivity:

T	RH	ETHYLENE PROD.	SENS.	CHILLING INJURIES
★★	★★	★	★★★	-

Heat production:

HEAT PRODUCTION
W/ton

500

400

300

200

100

0   5   10   15   20   °C
TEMPERATURE, °C

Specific heat: 3.80 kJ/kg x °C 0.91 Btu/lb x °F
Specific weight: Palletized boxes 250-300 kg/m³. Bulk approx. 500 kg/m³

	Latin:	Brassica oleracea L. convar. capitata (L.) Alef. var. conica DC.
CABBAGE, HEAD POINTED	French:	Chou cabus de printemps or Chou pointu
	German:	Spitzkohl
	Spanish:	Col tipo corazón de buey or Col picuda

DESCRIPTION OF PRODUCT:

Cabbage is one of the oldest vegetables. The cultivated cabbage originates from wild cabbage which originally grew in Asia Minor and along the European coastal regions of the Mediterranean Sea and the Atlantic Ocean. The earliest cultivated varieties were loose-leaved vegetables.

Headed cabbages are presumed to have come into existance in the 9th century, but the first types had loosely closed heads unlike those firm heads we know today.

Pointed cabbage and white cabbage have a great deal in common historically, the only difference being the shape of the head. In many countries this type is regarded as an early variety of white cabbage, i.e. a cabbage with a short growth period.

Pointed cabbage may present some difficulties in cultivation as harvest time is critical. If it is cut too early the head is not compact and if it is harvested a couple of days too late the flower stalk may split or burst the head.

MAJOR PRODUCERS:

The major producers of cabbage in general are U.S.S.R., China, South Korea, Japan and Poland. Since pointed cabbage is often regarded as only an early variety no specific information on production is available.

STANDARDS:

For trade within the EEC pointed cabbages must comply with EEC Standard No. 24. There are many recommendations, e.g. U.S. Grade Standards, but they are not mandatory in international trade.

MINIMUM REQUIREMENTS:

Pointed cabbage should be intact, without cracks, clean and free from adhering soil, insects, rot and mould. They should expel no foreign smell or taste. The heads should be reasonably firm without discolouration of the outer or inner leaves and without signs of stem growth on the head. The head should be cut off directly below the lowest leaf.

Latin:	Brassica oleracea L. convar. capitata (L.) Alef. var. conica DC.
French:	Chou cabus de printemps or Chou pointu
German:	Spitzkohl
Spanish:	Col tipo corazón de buey or Col picuda

HEAD POINTED **CABBAGE**

KEEPING QUALITIES:
Shelf life:

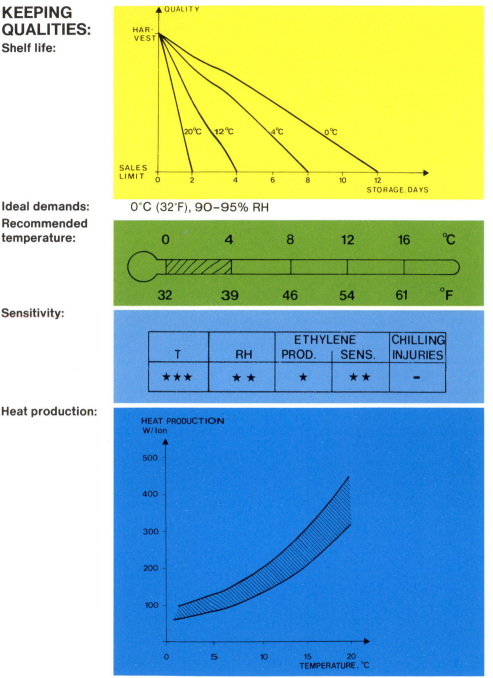

Ideal demands: 0°C (32°F), 90–95% RH

Recommended temperature:

		ETHYLENE		CHILLING
T	RH	PROD.	SENS.	INJURIES
★★★	★★	★	★★	–

Sensitivity:

Heat production:

Specific heat: 3.93 kJ/kg x °C 0.94 Btu/lb x °F
Specific weight: Boxes, pallets 200-300 kg/m³. Bulk approx. 500 kg/m³

	Latin:	Brassica oleracea L. convar. capitata (L.)
		Alef. var. rubra DC
	French:	Chou rouge
CABBAGE, RED	German:	Rotkohl
	Spanish:	Col roja or Lambarda

DESCRIPTION OF PRODUCT:

Cabbages rank among the oldest vegetables and originate as a cultivated plant from Asia Minor. The edible part on early cabbage types consisted only of spread leaves. The head-forming varieties are believed to have come much later.

The red cabbage is, like other cabbages, a biennial whose closely set leaves - the head - contain the plant's nutrients stored during the first year. The plant uses these nutrients during the second year for the building up of its inflorescence which shoots from the stem tip in the centre of the cabbage head.

The red pigments in red cabbage lie in the leaves' epidermal cells and the outermost layer of cells, whilst the inner parts are white. The intensity of the colour is an essential criterion in the hybridization of new red cabbage varieties. Red cabbage varieties are divided into groups according to their earliness, summer red cabbage, autumn red cabbage and hardy winter red cabbage. The winter red cabbage contains more pigments than the earlier varieties.

MAJOR PRODUCERS:

The major producers of red cabbage are West Germany, the Netherlands, Poland and the Scandinavian countries.

STANDARDS:

For trade within the EEC red cabbages must comply with EEC Standard No. 24. There are many recommendations, e.g. U.S. Grade Standards and OECD Standards, but they are not mandatory in international trade.

MINIMUM REQUIREMENTS:

Red cabbages should be intact, look fresh and have no growth cracks. The cabbage should have a firm head which is free from adhering soil, insects, rot and mould and from frost and mechanical damage. The stem should be cut off directly below the lowest leaf. The cabbage should not emit any foreign smell or taste.

Latin:	Brassica oleracea L. convar. capitata (L.) Alef. var. rubra DC		
French:	Chou rouge		
German:	Rotkohl		RED **CABBAGE**
Spanish:	Col roja or Lambarda		

KEEPING QUALITIES:

Shelf life:

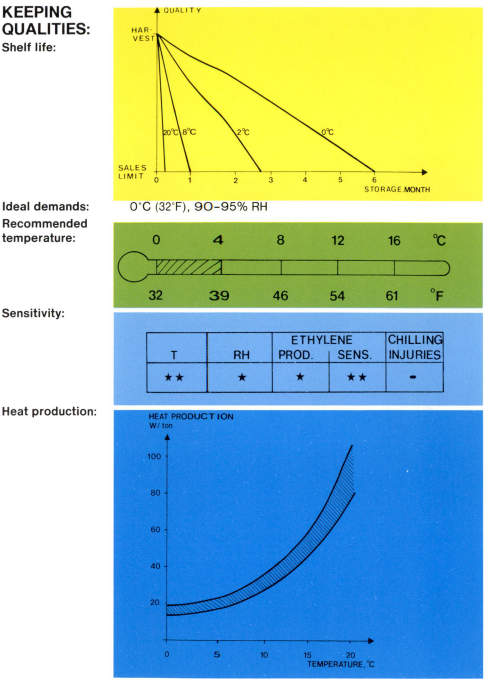

Ideal demands: 0°C (32°F), 90–95% RH

Recommended temperature:

Sensitivity:

T	RH	ETHYLENE PROD.	SENS.	CHILLING INJURIES
★★	★	★	★★	-

Heat production:

Specific heat: 3.93 kJ/kg x °C 0.94 Btu/lb x °F

Specific weight: Palletized boxes 200-300 kg/m³. Bulk approx. 550 kg/m³

CABBAGE, WHITE

Latin:	Brassica oleracca L. convar. capitata (L.) Alef. var. alba DC.
French:	Chou blanc
German:	Weisskohl
Spanish:	Col repollo

DESCRIPTION OF PRODUCT:

Wild cabbage can be found in many places in the Mediterranean region and in Southern England. The cultivated varieties came originally from Asia Minor. The compact and heavy heads, as we know them today, came into existence in the 16th century.

The cabbage plant is a biennial which has tightly overlapping leaves that form the head. This head contains food reserves accumulated during the first year. These reserves are used in the second year for the development of the inflorescence that shoots up from the tip of the thick stem inside the head.

There are many varieties of cabbage in various colours, shapes and sizes. A white cabbage normally weighs between 1 and 3 kg.

MAJOR PRODUCERS:

The major producers of cabbage are U.S.S.R., China, South Korea, Japan and Poland.

STANDARDS:

For trade within the EEC, white cabbage must comply with EEC Standard No. 24. Besides this there are many recommendations, e.g. U.S. Grade Standards, but they are not mandatory in international trade.

MINIMUM REQUIREMENTS:

White cabbage should be intact, free of cracks, rot and mould, adhering soil and insects. There should be no foreign smell or taste. The head should be firm and compact and without discolouration of the inner or outer leaves. The stem should be cut off directly below the lowest leaf.

Latin: Brassica oleracca L. convar. capitata (L.)
ef. var. alba DC.
French: Chou blanc
German: Weisskohl
Spanish: Col repollo

WHITE **CABBAGE**

KEEPING QUALITIES:
Shelf life:

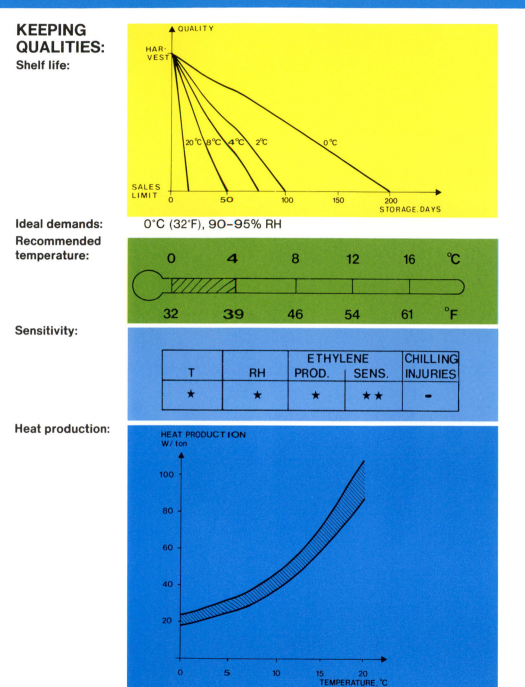

Ideal demands: 0°C (32°F), 90–95% RH

Recommended temperature:

0	4	8	12	16	°C
32	39	46	54	61	°F

Sensitivity:

T	RH	ETHYLENE PROD.	SENS.	CHILLING INJURIES
★	★	★	★★	-

Heat production:

Specific heat: 3.93 kJ/kg x °C 0.94 Btu/lb x °F

Specific weight: Palletized boxes 200-300 kg/m³. Bulk approx. 550 kg/m³

GOOSEBERRY, CAPE

Latin:	Physalis peruviana L. var. edulis
French:	Alkekenge, physalis
German:	Ananaskirsche, Kapstachelbeere
Spanish:	Uvilla

DESCRIPTION OF PRODUCT:

Cape gooseberries originate from Central and South America. The plant, which is a climber, was cultivated by the Incas for its health-giving properties (high vitamin content).

Cultivation of cape gooseberries spread along the major shipping routes for use in prevention of scurvy. They are now grown almost all over the world where the climate is suitable.

Cape gooseberries are greenish-yellow, cherry-sized berries with a smooth skin. The juicy flesh is yellowish and contains small, soft seeds. Its sweet-sour taste resembles that of gooseberries. Each berry lies in a light brown capsule (an enlarged flower pod).

MAJOR PRODUCERS:

The major producers of cape gooseberries are Kenya, Colombia, India, South Africa and Mexico.

STANDARDS:

There are no international standards for cape gooseberries.

MINIMUM REQUIREMENTS:

Cape gooseberries should comprise a wrap or capsule with a berry in it. The wrap should be undamaged, free from spots and have a uniform golden colour. The berries should be yellow approaching orange and they should be juicy and flavourful without any after-taste. There should be no signs of attack by disease or pests.

Latin: Physalis peruviana L. var. edulis
French: Alkekenge, physalis
German: Ananaskirsche, Kapstachelbeere
Spanish: Uvilla

CAPE **GOOSEBERRY**

KEEPING QUALITIES:

Shelf life:

14°C, 80% RH, 1-2 months
20°C, 60% RH, 1-2 weeks

Ideal demands: 12-15°C (54-59°F), 80% RH

Recommended temperature:

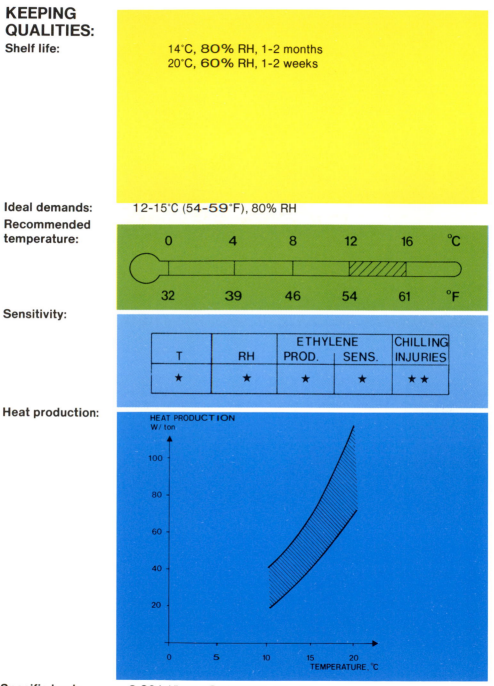

	°C
0 4 8 12 16	°C
32 39 46 54 61	°F

Sensitivity:

T	RH	ETHYLENE PROD.	ETHYLENE SENS.	CHILLING INJURIES
★	★	★	★	★★

Heat production:

HEAT PRODUCTION
W/ ton

TEMPERATURE, °C

Specific heat: 3.60 kJ/kg x °C 0.86 Btu/lb x °F
Specific weight: Boxes, pallets 150-200 kg/m³

83

(Star-apple, star-fruit)
CARAMBOLA

Latin:	Averrhoa carambola L.
French:	Carambolier or Averrhoa
German:	Karambole or Sternfrucht
Spanish:	Carambola

DESCRIPTION OF PRODUCT:

The carambola is said to have its origin in tropical Asia and it still grows wild in Indonesia. It spread from there to the rest of the tropical world and sub-tropical areas where exposure to night frost is least likely. The tree, an evergreen, is small and very decorative due to its pink blossoms.

The carambola is an oval, yellow fruit with lengthwise ridges which give the fruit a star-shaped cross-section. The flesh is light yellow and translucent. Its taste is either sour or sourish, depending on the variety and stage of ripeness.

The carambola contains oxalic acid and has a considerable content of Vitamin C. The ridged edges of the fruit are susceptible to bruising which will result in brown discolouration.

MAJOR PRODUCERS:

The major producers of carambola are Thailand, Brazil and Israel.

STANDARDS:

There are no international standards for carambola.

MINIMUM REQUIREMENTS:

Carambola should be fresh, intact, clean and sound. They should be full-bodied and free of spots, brown edges and other injuries. The colour should be yellow or greenish yellow with a wax-like almost transparent look. Depending on the variety, the fruit should have a fresh, sourish or sweet-sour taste without being unpleasant. There should be no signs of attack by disease or pests.

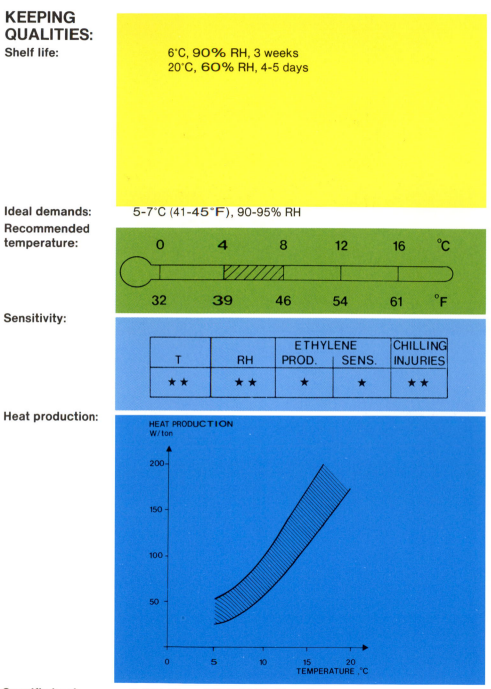

	Latin:	Averrhoa carambola L.
	French:	Carambolier or Averrhoa
	German:	Karambole or Sternfrucht
	Spanish:	Carambola

(Star-apple, star-fruit)
CARAMBOLA

KEEPING QUALITIES:

Shelf life:

6°C, **90%** RH, 3 weeks
20°C, **60%** RH, 4-5 days

Ideal demands: 5-7°C (41-45°F), 90-95% RH

Recommended temperature:

°C	0	4	8	12	16
°F	32	39	46	54	61

Sensitivity:

T	RH	ETHYLENE PROD.	SENS.	CHILLING INJURIES
★ ★	★ ★	★	★	★ ★

Heat production:

HEAT PRODUCTION
W/ton

Specific heat: 3.89 kJ/kg x °C 0.93 Btu/lb x °F
Specific weight: Cartons, pallets 250-300 kg/m³

Latin:	Daucus carota L.
French:	Carotte
German:	Möhre
Spanish:	Zanahoria

CARROT

DESCRIPTION OF PRODUCT:
Carrots came originally from Asia Minor.
The plant is a biennial and during the
first year it develops a strong root which
is the edible part. In the second year it
seeds and the root becomes woody and
unfit for consumption. The shape and
size depend on the variety. The cylindrical or slightly conical varieties are most
common, but there are also round and
other in-between shapes.

MAJOR PRODUCERS:
The major producers of carrots are China, U.S.S.R., U.S.A., Poland and Japan.

STANDARDS:
For trade within the EEC carrots have to
comply with EEC Standard No. 17. There
are several other recommendations, e.g.
U.S. Grade Standards, but they are not
mandatory in international trade.

MINIMUM REQUIREMENTS:
Carrots should be intact, sound, clean,
free from attack by disease, pests,
mould or rot and without foreign smell or
taste. Carrots should not have cracks
splits or a woody texture. There should
be no foreign matter or remnants from
tops on the roots. The colour should
be yellow-orange without any green or
other discolouration.

KEEPING QUALITIES:

Shelf life:

Winter carrots.

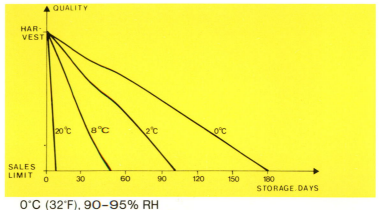

Ideal demands: 0°C (32°F), 90–95% RH

Recommended temperature:

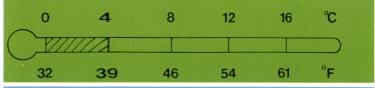

Sensitivity:

T	RH	ETHYLENE PROD.	SENS.	CHILLING INJURIES
★★	★★	★	★★	–

Heat production:

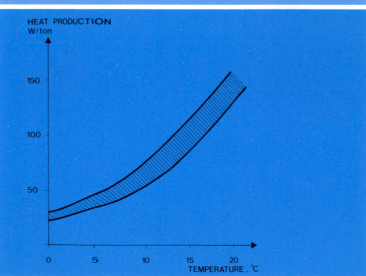

Specific heat: 3.92 kJ/kg x °C 0.94 Btu/lb x °F

Specific weight: Palletized crates 380-400 kg/m³. Bulk approx. 440 kg/m³

87

	Latin:	Brassica oleracea L. convar. botrytis (L.)
CAULIFLOWER		Alef. var. botrytis L.
	French:	Chou-fleur
	German:	Blumenkohl
	Spanish:	Coliflor

DESCRIPTION OF PRODUCT:

The ancient forms of cultivated cauli-flower originate from Asia Minor. The plant is cultivated as an annual. The head of the cauliflower comprises mainly barren inflorescence full of nutrient reserves. The colour is white or light cream and its diameter is normally around 10-20 cm. The flower head is sensitive to rough handling. Cauliflowers are usually harvested with the surrounding leaves serving as protection against damage to the head.

MAJOR PRODUCERS:

The major producers of cauliflower are China, India, Italy, France and U.K.

STANDARDS:

For trade within the EEC cauliflowers must comply with EEC Standard No. 3. Additionally, there are many recommendations, e.g. U.S. Grade Standards, but they are not mandatory in international trade.

MINIMUM REQUIREMENTS:

Cauliflowers should be intact, clean, sound and fresh, free of any foreign matter and pests and not affected by disease. There should be no trace of any foreign smell or taste. The head should be pale, without discolouration and the attached leaves green and fresh. There should be no signs of wilting or mechanical injuries.

Latin: Brassica oleracea L. convar. botrytis (L.)
Alef. var. botrytis L.
French: Chou-fleur
German: Blumenkohl
Spanish: Coliflor

CAULIFLOWER

KEEPING QUALITIES:

Shelf life:

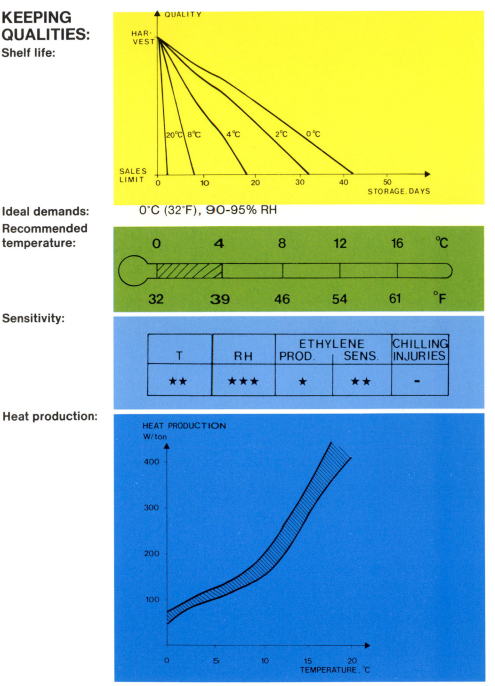

Ideal demands: 0°C (32°F), 90-95% RH

Recommended temperature:

Sensitivity:

	T	RH	ETHYLENE PROD.	SENS.	CHILLING INJURIES
	★★	★★★	★	★★	-

Heat production:

Specific heat: 4.02 kJ/kg x °C 0.96 Btu/lb x °F

Specific weight: Palletized boxes 200-230 kg/m³. Bulk approx. 320 kg/m³

89

CELERIAC

Latin:	Apium graveolens L. var. rapaceum (Miller) Gaudich
French:	Céleri rave
German:	Knollensellerie
Spanish:	Apio nabo

DESCRIPTION OF PRODUCT:

Celeriac came originally from the Mediterranean region. It has been known for several thousand years in North Africa and Greece.

Celeriac is a biennial. The root swells and forms a tuber that is almost spherical. It has typical diameter of 10-15 cm and weighs 0.5-1 kg.

The root has a thin, brown skin. The flesh is pale, white or yellowish. Celeriac is closely related to celery. Only in the 16th century did they appear as 2 different types.

should neither be hollow nor the flesh discoloured. The tubers should only bear a few roots and be free of shoots. If celeriac is sold with leaves, these should be fresh and green.

MAJOR PRODUCERS:

The major producers of celeriac are in Europe.

STANDARDS:

There are no international standards for celeriac.

MINIMUM REQUIREMENTS:

Celeriac should be firm, free from adhering soil, rot, mould and pests. The tubers

Latin:	Apium graveolens L. var. rapaceum
	(Miller) Gaudich
French:	Céleri rave
German:	Knollensellerie
Spanish:	Apio nabo

CELERIAC

KEEPING QUALITIES:
Shelf life:

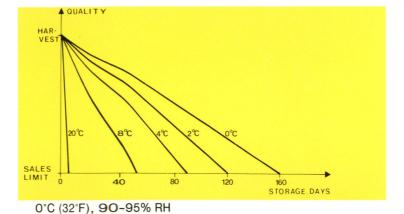

Ideal demands: 0°C (32°F), 90–95% RH

Recommended temperature:

0	4	8	12	16	°C
32	39	46	54	61	°F

Sensitivity:

T	RH	ETHYLENE PROD.	SENS.	CHILLING INJURIES
★	★	-	-	-

Heat production:

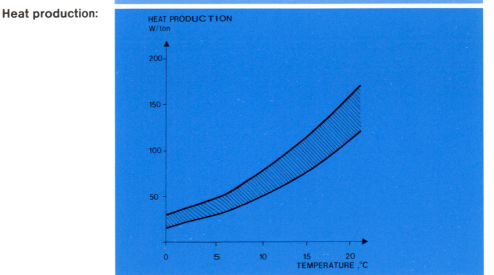

Specific heat: 3.87 kJ/kg x °C 0.92 Btu/lb x °F

Specificweight: Boxes, pallets 300-350 kg/m³. Bulk approx. 525 kg/m³

91

DESCRIPTION OF PRODUCT:

Celery, a biennial, is closely related to celeriac and is found in its wild form in many parts of the world.

Celery was cultivated in ancient times in Egypt and Greece for medicinal purposes. These types were unpleasant in smell and taste. In the 16th century in Italy the plant was improved to a milder and more pleasant tasting variety. In 1806 celery appeared for the first time in the U.S.A. It spread mainly to the English speaking countries.

Celery does not develop tubers, but long, sturdy, upright leaf stalks which are almost colourless in some varieties.

The crisp, fleshy stalks form the edible part of the plant. They are used cooked or raw in salads.

MAJOR PRODUCERS:

The major producers of celery are the Netherlands, Belgium, U.S.A., Israel and U.K.

STANDARDS:

For trade within the EEC celery must comply with EEC Standard No. 26. There are many recommendations, e.g. U.S. Grade Standards, which are not mandatory in international trade.

MINIMUM REQUIREMENTS:

Celery should be intact (though the top may be cut off), clean, fresh, sound, disease and pest-free, not running to seed, free of shoots and sunken parts. There should be no foreign smell or taste. The remaining part of the root should not exceed 5 cm.

Celery should be fully developed, regular in shape and free from damage, cracks or coarse fibres.

KEEPING QUALITIES:
Shelf life:

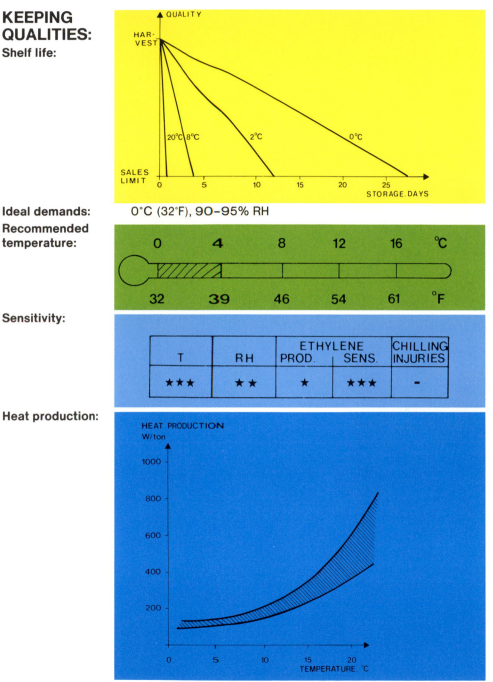

Ideal demands: 0°C (32°F), 90–95% RH

Recommended temperature:

0	4	8	12	16 °C
32	39	46	54	61 °F

Sensitivity:

T	RH	ETHYLENE PROD.	SENS.	CHILLING INJURIES
★★★	★★	★	★★★	–

Heat production:

Specific heat: 3.97 kJ/kg x °C 0.95 Btu/lb x °F
Specificweight: Palletized cartons 100–150 kg/m³. Bulk approx. 200 kg/m³

93

CHANTERELLE

Latin:	Cantharellus sp.
French:	Cantharella
German:	Pfifferling
Spanish:	Cantarela

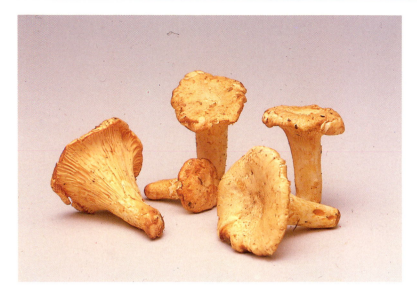

DESCRIPTION OF PRODUCT:

Chanterelle is a common fungus in forest regions, particularly in Europe. Chanterelle has a funnel-shaped fruiting body along the undersurface of which are the gills where the spores are produced.

The young fungus is arched and the edge of the cup curls up. It later becomes funnel-shaped, the cap rolls out and ends with a down-curled edge. Chanterelle has a more distinctive taste than mushroom. Fresh chanterelle are egg yolk-yellow. They are also available dried, but these are dark.

Chanterelle cannot be cultivated and, therefore, those that are on the market are gathered from the wild.

MAJOR PRODUCERS:

Chanterelle is not produced under cultivation. They are gathered in forest regions all over Europe.

STANDARDS:

There are no international standards for chanterelle.

MINIMUM REQUIREMENTS:

Chanterelle should be sound, adequately mature and have its characteristic shade of yellow. They should be free from discolouration, pressure marks or damage. They should not be tough or woody. Chanterelle should not be infected by disease or pests nor have any off-taste.

Latin: Cantharellus sp.
French: Cantharella
German: Pfifferling
Spanish: Cantarela

CHANTARELLE

KEEPING QUALITIES:

Shelf life:

0°C, **90%** RH, 2 weeks
20°C, **60%** RH, 2-3 days

Ideal demands: 0-1°C (32-34°F), 85-90% RH

Recommended temperature:

0	4	8	12	16	°C
32	39	46	54	61	°F

Sensitivity:

T	RH	ETHYLENE PROD.	SENS.	CHILLING INJURIES
★★	★★	–	–	–

Heat production:

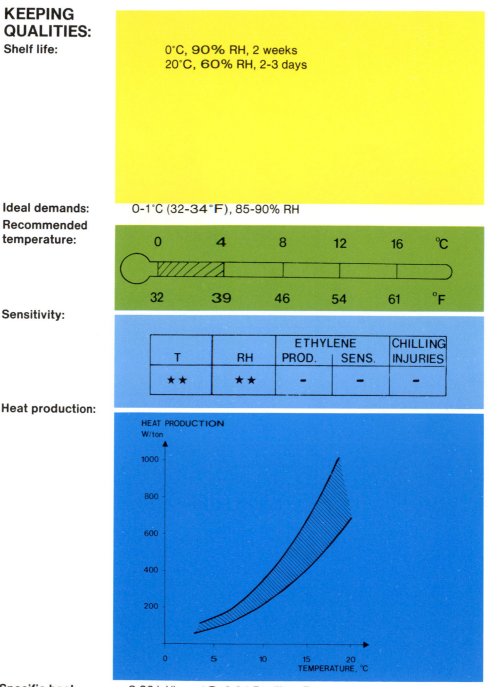

HEAT PRODUCTION
W/ton

TEMPERATURE, °C

Specific heat: 3.93 kJ/kg x °C 0.94 Btu/lb x °F
Specific weight: Boxes, pallets 220-280 kg/m3

DESCRIPTION OF PRODUCT:

Cherimoya originally came from Peru where it had been cultivated for several thousand years. The name means »cold seeds» in Indian and refers to the fact that cherimoya was cultivated on tropical highlands, at an altitude of 1000-2000 m. After the discovery of America cherimoya cultivation spread to other areas, including Mediterranean region. It is now cultivated in many parts of the world including the subtropical regions.

The cherimoya tree is of average height (5-6m). Its flowers are self-fertile. Each cherimoya consists of many small fruits that have grown together, thus giving the surface a scaly appearance.
The pulp of the fruit comprises partly the small fruits and partly the expanded receptacle.
The cherimoya is oval or heart-shaped. The skin is green with finger-like marks on the surface. The pulp is pale and soft and contains many dark seeds. The fruit is ripe when the pulp yields to slight pressure. The pulp must be white or slightly greyish and its consistency cream-like. As different varieties have various skin colours it may be difficult to judge the fruit's ripeness by the colour.

MAJOR PRODUCERS:

The major producers of cherimoya are Venezuela, Colombia, Dominican Republic, Thailand and Spain.

STANDARDS:

There are no international standards for cherimoya.

MINIMUM REQUIREMENTS:

Cherimoya should be intact, sound and without any foreign smell or taste. The skin should be free of cracks and mechanical damage. There should be no visible signs of attack by disease or pests. Rot is very common, usually starting at the stem end, but should only be present to a minor degree. The pulp of chilling-injured cherimoya discolours and tends to rot.

KEEPING QUALITIES:

Shelf life:

12°C, 90% RH, 2-3 weeks
20°C, 60% RH, 3-4 days

Ideal demands: 12-14°C (54-57°F), 85-90% RH

Recommended temperature:

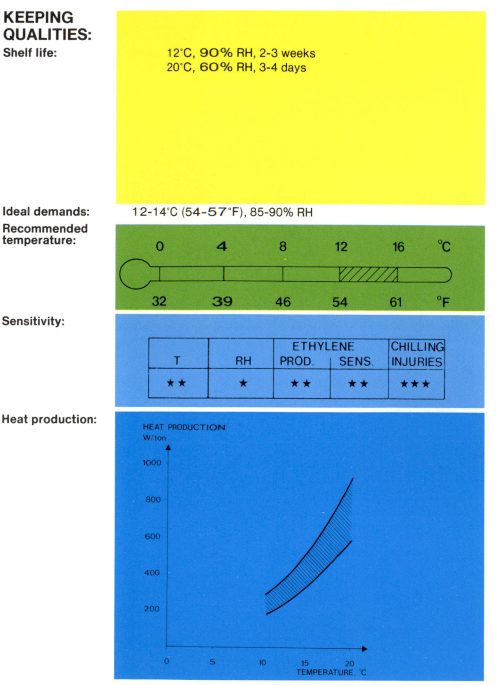

0	4	8	12	16	°C
32	39	46	54	61	°F

Sensitivity:

T	RH	ETHYLENE PROD.	SENS.	CHILLING INJURIES
★★	★	★★	★★	★★★

Heat production:

HEAT PRODUCTION
W/ton

TEMPERATURE, °C

Specific heat: 3.35 kJ/kg x °C 0.80 Btu/lb x °F
Specific weight: Palletized cartons 220-280 kg/m³

Latin:	Prunus cerasus L. and Prunus avium L.
French:	Cerise
German:	Kirsche
Spanish:	Cereza

DESCRIPTION OF PRODUCT:

Cherries have been cultivated in China for 4000 years. They belong to the rose family (Rosaceae) and comprise two types, sweet cherry (Prunus cerasus) and sour cherry (Prunus avium).
It is presumed that sweet cherry originally came from regions around the Black Sea and Caspian Sea and the sour from the region that lies between Switzerland and the Adriatic Sea.

The sweet cherry tree, is larger than the sour, and is self-sterile. It is, therefore, necessary to plant different varieties so as to ensure pollination.
Sour cherry trees are self-fertile and may be grown singly.
Both light and dark varieties of sweet cherry are cultivated, but only dark varieties of the sour cherry are grown.

There are over 600 varieties of cherries. The colour depends on the variety. The berries are normally 15-25 mm in diameter with juicy flesh and a small, round stone.

MAJOR PRODUCERS:

Cherries are produced in many parts of the world with North America, Europe, China, Japan and Australia as major producers.

STANDARDS:

For trade within the EEC cherries must comply with EEC Standard No. 10. There are many recommendations, e.g. U.S. Grade Standards, but they are not mandatory in international trade.

MINIMUM REQUIREMENTS:

Cherries should be intact, sound and firm. They should appear fresh, be clean and free from attack by pests and from any foreign taste or smell. The major part of the crop should bear stalks and berries should be typical of the variety for colour, shape and development. Cherries should be free from sun, hail or mechanical damage, spots, scratches, bruises and cracks.

KEEPING QUALITIES:

Shelf life:

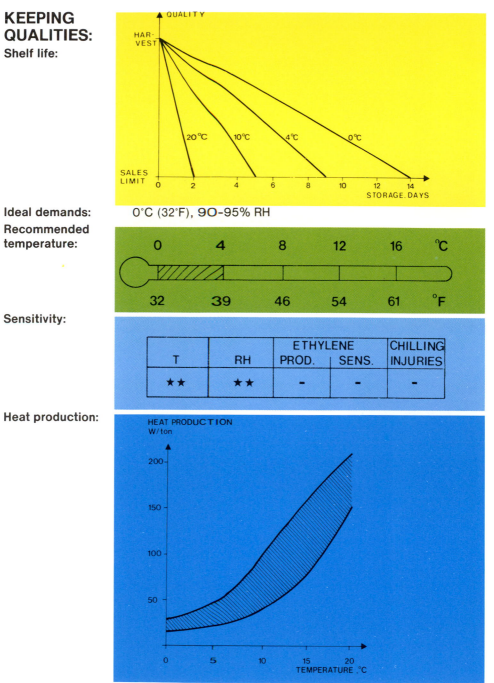

Ideal demands: 0°C (32°F), 90-95% RH

Recommended temperature:

Sensitivity:

T	RH	ETHYLENE PROD.	SENS.	CHILLING INJURIES
★★	★★	-	-	-

Heat production:

Specific heat: 3.64 kJ/kg x °C 0.87 Btu/lb x °F

Specific weight: Boxes, pallets 350-400 kg/m^3. Bulk approx. 700 kg/m^3

Latin:	Cichorium intybus L. var. foliosum Hegi
French:	Chicorie-Witloof or endive
German:	Zichoriensalat, Chicoree or Witloof
Spanish:	Achicoria de barba gruesa

CHICORY (Witloof)

DESCRIPTION OF PRODUCT:

Cultivated chicory originates from the wild chicory that comes from the Mediterranean region.

There are two phases in the cultivation of chicory. Firstly, outdoor culture for the production of sturdy taproots. These roots are then placed side by side for further growth in dark greenhouses. The roots begin to shoot and these shoots (chicory) are cut off when they reach the size desired. Cultivation in greenhouses lasts approx. 3 weeks. This is a demanding process and explains why chicory is more expensive than other types of lettuce.

Chicory must not receive light which will cause the leaves to turn green. Green leaves are bitter.

MAJOR PRODUCERS:

The major producers of chicory are France, Belgium and the Netherlands.

STANDARDS:

For trade within the EEC chicory must comply with EEC Standard No. 12. There are many recommendations, e.g. U.S. Grade Standards, but they are not mandatory in international trade.

MINIMUM REQUIREMENTS:

Chicory should be whole, sound, fresh, clean and free from disease and attack by pests. The heads should show no signs of bolting. All roots should be removed. The heads should be free of discolouration, frost and mechanical damage. Chicory should be pale with no greening of the leaves and free from any foreign smell or taste. The heads should be full-bodied and firm and should not be open.

During storage, transport and even in retail display chicory should be kept out of sun light as even after harvest chicory will revert to a green colour.

Latin:	Cichorium intybus L. var. foliosum Hegi
French:	Chicorie-Witloof or endive
German:	Zichoriensalat, Chicoree or Witloof
Spanish:	Achicoria de barba gruesa

(Witloof) **CHICORY**

KEEPING QUALITIES:

Shelf life:

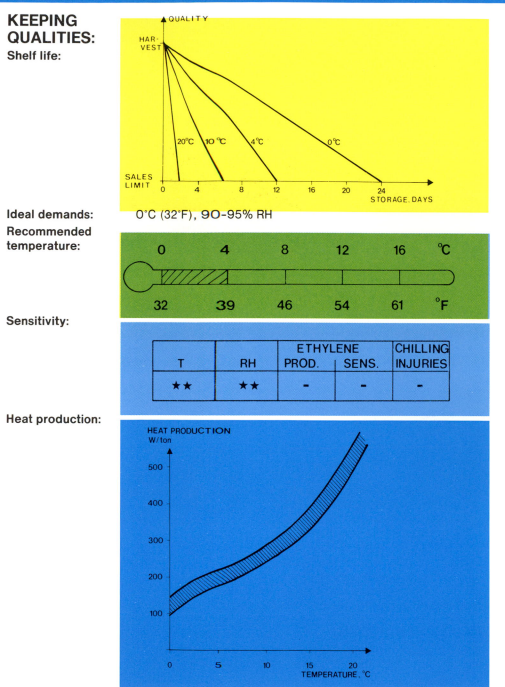

Ideal demands: 0°C (32°F), 90-95% RH

Recommended temperature:

Sensitivity:

T	RH	ETHYLENE PROD.	SENS.	CHILLING INJURIES
★★	★★	–	–	–

Heat production:

Specific heat: 4.01 kJ/kg x °C 0.96 Btu/lb x °F

Specific weight: Boxes, pallets 300-380 kg/m³. Bulk approx. 450 kg/m³

101

CHILLI (Hot pepper)

Latin: Capsicum sp. L.
French: Piment fort du chili
German: Peperoni
Spanish: Chile

DESCRIPTION OF PRODUCT:

Chillies came originally from South America. The annual plant is approx. 1.5 m tall with upright branched growth. The fruit normally has a tapering form and can vary in length, colour and pungency. All unripe chillies are green, but during the ripening process they turn red or reddish yellow. The fruit is hollow and contains many seeds.

In common with other types of capsicum, chillies contain capsaicin in the seeds and walls of the fruit. It is this compound that gives the fruit its characteristic hot taste. As all types of peppers do not contain the same amount of capsaicin, the degree of pungency may vary from one variety to another. The smallest varieties are usually the hottest.

MAJOR PRODUCERS:

The major producers of chillies are China, Nigeria, Turkey, Mexico and Indonesia.

STANDARDS:

There are no international standards for chillies.

MINIMUM REQUIREMENTS:

Chillies should be intact, fresh, clean and sound. They should be fully developed and free from injury and disease or pests. The colour should be even and the fruit firm and full-bodied. The most common defects are injury, rot and signs of wilting.

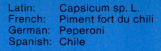

Latin:	Capsicum sp. L.
French:	Piment fort du chili
German:	Peperoni
Spanish:	Chile

(Hot pepper) **CHILLI**

KEEPING QUALITIES:

Shelf life:

10°C, 90% RH, 2-3 weeks
20°C, 60% RH, 2-3 days

Ideal demands: 8-10°C (46-50°F), 90-95% RH

Recommended temperature:

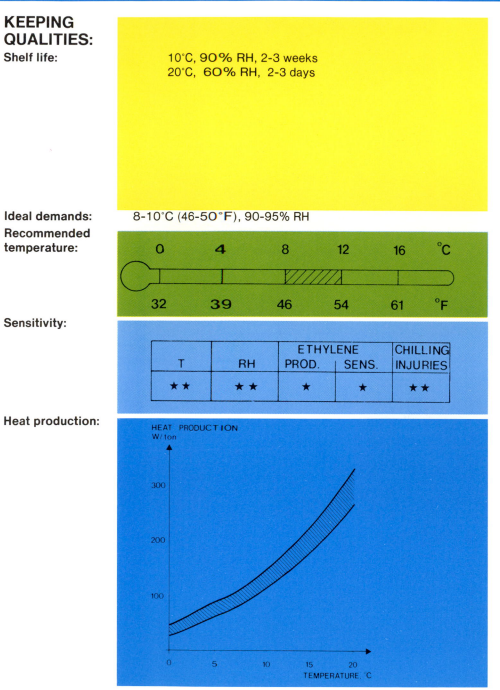

Sensitivity:

T	RH	ETHYLENE PROD.	SENS.	CHILLING INJURIES
★★	★★	★	★	★★

Heat production:

Specific heat: 3.98 kJ/kg x °C 0.95 Btu/lb x °F
Specific weight: Boxes, pallets 180-220 kg/m³. Bulk approx. 300 kg/m³

ARTICHOKE, CHINESE

Latin: Stachys sieboldi Miq.
French: Crosne or Crosnes du Japon
German: Crosne or Chinesische Artischoke
Spanish: Estaquide or Alcachofa tuberosa

DESCRIPTION OF PRODUCT:

Chinese artichoke originates from East Asia, presumably from Japan.

It is an herbaceous plant that develops a considerable number of underground stolons whose last 5 cm swell up to form whitish or brownish spherical tubers. The tubers resemble small screw-shaped Jerusalem artichoke and they are skinless and white inside.

Their carbohydrate content is not starch, but an insulin-like sugar compound.

MAJOR PRODUCERS:

The major producers of Chinese artichokes are China, Japan, France and Belgium.

STANDARDS:

There are no international standards for Chinese artichokes.

MINIMUM REQUIREMENTS:

Chinese artichokes should be intact, clean, sound and free from attack by disease or insects, mechanical damage, and broken or wilted tubers.

Latin: Stachys sieboldi Miq.
French: Crosne or Crosnes du Japon
German: Crosne or Chinesische Artischoke
Spanish: Estaquide or Alcachofa tuberosa

CHINESE **ARTICHOKE**

KEEPING QUALITIES:

Shelf life:

0°C, **95%** RH, 1-2 weeks
20°C, **60%** RH, 1-2 days

Ideal demands: 0°C (32°F), **90**-95% RH

Recommended temperature:

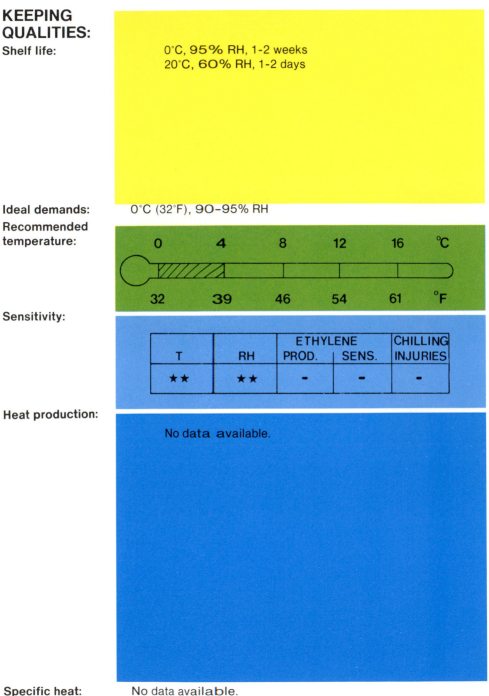

| | 0 | 4 | 8 | 12 | 16 | °C |
| 32 | 39 | 46 | 54 | 61 | °F |

Sensitivity:

T	RH	ETHYLENE PROD.	SENS.	CHILLING INJURIES
★★	★★	-	-	-

Heat production:

No data available.

Specific heat: No data available.

Specific weight: Boxes, pallets 300-380 kg/m³. Bulk approx. 450 kg/m³

105

CABBAGE, CHINESE

DESCRIPTION OF PRODUCT:

Chinese cabbage originally came from East Asia including China, where it has been cultivated for a very long time. From there it spread to the other regions of Asia, especially Japan and Taiwan, where Chinese cabbage is now widespread. Large scale cultivation of this cabbage in other continents did not begin until recently.

Chinese cabbage is an annual in which the rosette-formed tuft of leaves grows into a fairly compact head. The head houses food reserves for the inflorescence which grows from the tip of the stem.

There are many different types of Chinese cabbage, from long, slender heads where the leaf-stalks are very broad and heads that are open to very rounded heads with crimpled, light green leaves that have narrow ribs and closed heads.

MAJOR PRODUCERS:

The major producers of Chinese cabbage are China, Japan, Korea, Austria and Spain.

STANDARDS:

There are no international standards for Chinese cabbage.

MINIMUM REQUIREMENTS:

Chinese cabbage should be intact, fresh, sound and full-bodied with no signs of disease or mechanical damage. The cabbage should neither have dark vascular bundles nor any form of discolouration or spots. They should be cut directly below the lowest leaf. The cut should be clean and dry. Chinese cabbage should not show signs of shoot growth or contain visible bolters. They should be free from foreign smell or taste.

Latin: Brassica pekinensis L.
French: Chou de Chine
German: Chinakohl
Spanish: Col chino

KEEPING QUALITIES:

Shelf life:
Chinese cabbage may be stored longer, but they shall need trimming.

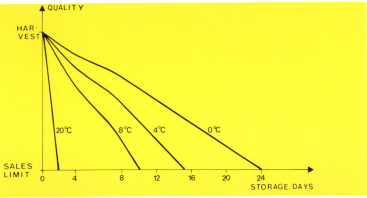

Ideal demands: 0-4°C (32-39°F), 90-95% RH, depending on the variety.

Recommended temperature:

0	4	8	12	16	°C
32	39	46	54	61	°F

Sensitivity:

T	RH	ETHYLENE PROD.	SENS.	CHILLING INJURIES
★★	★★	–	★★	★

Heat production:

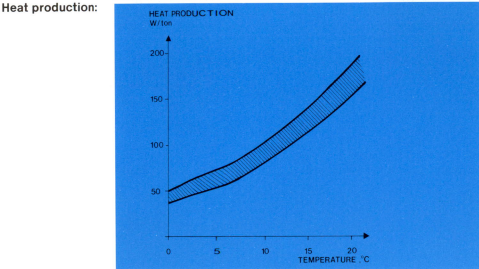

Specific heat: 4.04 kJ/kg x °C 0.97 Btu/lb x °F
Specific weight: Palletized boxes 220-250 kg/m³. Bulk approx. 250 kg/m³

107

DESCRIPTION OF PRODUCT:

There is an increasing number of citrus hybrids on the market. Three of them are mentioned here.

UGLI is a hybrid of orange, grapefruit and clementine. This fruit was found in Jamaica at the beginning of this century. Ugli is not widely grown outside Jamaica. Ugli is a large fruit with juicy, sweet and tasty flesh. The rind is thick, yellow and easy to peel.

MINNEOLA is a hybrid of clementine and grapefruit. It is an orange coloured fruit, a little larger than a mandarin. The rind is smooth, thin and easy to remove. The flesh, which resembles that of the grapefruit, is juicy, aromatic and sourish.

LIMEQUAT is a hybrid of kumquat and lime. There are other similar hybrids which primarily serve the purpose of producing miniature versions of citrus fruit. Limequats are yellowish green or green and are as big as kumquats. They taste like lime and are eaten with the rind.

MAJOR PRODUCERS:

The major producer of ugli is Jamaica, of minneola Israel and limequat Brazil, Israel and Italy.

STANDARDS:

Citrus fruits that are hybrids of mandarin, clementine or satsuma and have retained these fruits' characteristics must comply with EEC Standard No. 18 for trade within the EEC. Among the three fruits mentioned here minneola belongs to this category. There are no international standards for the other hybrids.

MINIMUM REQUIREMENTS:

The fruits should be intact, sound, free of frost damage and any visible foreign matter. They should also be free from cracks, spots, mechanical damage and signs of attack by disease or pests. They should not emit any foreign smell or taste other than that deriving from the use of statutory surface preservatives. Colour and shape should be typical of the variety.

KEEPING QUALITIES:

Shelf life:

Ugli, Minneola:	4°C, 90% RH, 3-5 weeks
	20°C, 60% RH, 7-10 days
Limequat:	10°C, 90% RH, 4 weeks
	20°C, 60% RH, 1 week

Ideal demands: 4-5°C (39-41°F), 90% RH, though 10°C (50°F) for limequat

Recommended temperature:

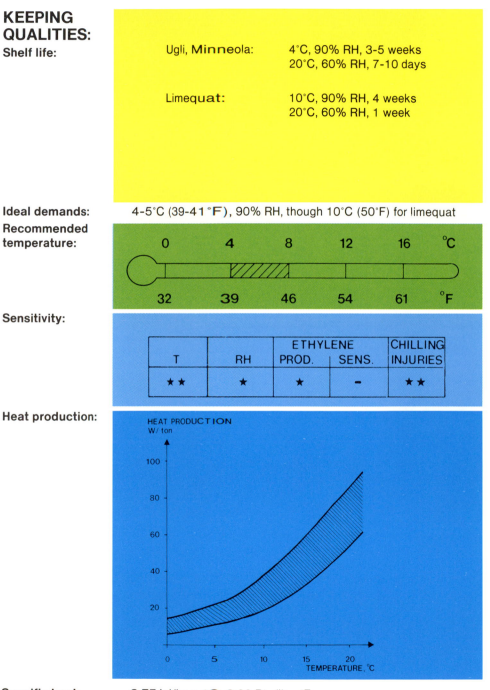

0	4	8	12	16	°C
32	39	46	54	61	°F

Sensitivity:

T	RH	ETHYLENE PROD.	SENS.	CHILLING INJURIES
★ ★	★	★	–	★ ★

Heat production:

HEAT PRODUCTION
W/ ton

TEMPERATURE, °C

Specific heat: 3.77 kJ/kg x °C 0.90 Btu/lb x °F
Specific weight: Palletized cartons 350-450 kg/m³. Bulk approx. 500 kg/m³

DESCRIPTION OF PRODUCT:

The term small citrus fruits covers a range of products that resemble the orange. They include different types of mandarin or hybrids between mandarin and other citrus species.

The original form of mandarin comes from Mauritius, an island off the east coast of Africa. Two other variants, the tangerine which originates from North Africa is named after the port of disembarkation -Tangier, and satsuma which comes from the Japanese island of Kyushu. Furthermore, there are numerous other hybrids between mandarin and other citrus species.

Temple is an American hybrid between grapefruit and mandarin. Further attempts at hybridizing the temple and the mandarin produced the Wilking variety which is cultivated solely in California.

Monreal, Ortanique and Tangelo are among other types that are available on the market. Small citrus fruits such as those mentioned above are widespread in all citrus-producing regions, but orga-nised cultivation started after the middle of the 19th century.

Apart from the size, the fruits differ from oranges by being sweeter in taste and stronger in aroma. The peel is thin and easy to remove.

MAJOR PRODUCERS:

The major producers of clementines, etc. are Japan, Spain, Italy, Brazil and U.S.A.

STANDARDS:

For trade within the EEC clementines, etc. must comply with EEC Standard No. 18. There are many recommendations, e.g. U.S. Grade Standards, but they are not mandatory in international trade.

MINIMUM REQUIREMENTS:

Clementines, etc. must be intact, sound and free from damage caused by frost and any visible foreign matter. The fruits must also be free from foreign smell or taste. The minimum requirement of juice content is 33% of the fruit's weight. Fruits that are affected by mould or rot should be excluded. The surface should be even and without any blemishes.

KEEPING QUALITIES:

Shelf life:

Clementines:	0-3°C, 1-2 weeks
Mandarines:	0-3°C, 8-10 weeks
Satsuma:	4°C, 8-12 weeks

At 20°C the keeping quality is approx. 3-7 days, depending on the variety.

Ideal demands: 0-4°C (32-39°F), 85-90% RH, depending on the variety.

Recommended temperature:

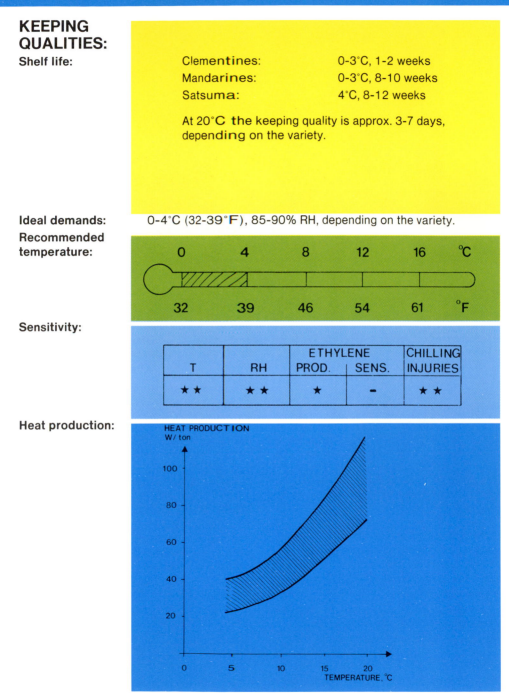

	T	RH	ETHYLENE PROD.	SENS.	CHILLING INJURIES
Sensitivity:	★ ★	★ ★	★	–	★ ★

Heat production:

HEAT PRODUCTION
W/ ton

(graph: heat production vs. temperature °C)

Specific heat: 3.77 kJ/kg x °C 0.90 Btu/lb x °F
Specific weight: Boxes, pallets 350-450 kg/m³. Bulk 460-550 kg/m³

(Summer squash, Zucchini)
COURGETTE

Latin: Cucurbita pepo L.
French: Patisson, Courgette or Courge d'été
German: Gemüse Kürbis, Courgette or Zucchini
Spanish: Calabazin

DESCRIPTION OF PRODUCT:

Courgettes originate from South America where they have been cultivated by the Indians for a very long time. The plant is an herbaceous annual vine with large, lobed leaves.

The flesh of the fruit is pale, slightly dry and embedded with numerous seeds. The courgette is an unripe marrow. If it is left to continue its growth on the plant it will develop into an ordinary marrow.

Courgettes are to be found in various colours and shapes. As a sales commodity the elongated, dark green type is preferred, but the round types also enjoy some popularity.

There are differing opinions as to when courgettes should be harvested. The size varies in length from 10 cm to approx. 30 cm and the weight from approx. 100 g to 500 g. If the fruit is left to grow too long the seedy pulp will coarsen and the decorative green part will diminish as the fruit gets bigger.

MAJOR PRODUCERS:

The major producers of courgettes are China, Rumania, Egypt, Argentina and Turkey.

STANDARDS:

For trade within the EEC courgettes must comply with EEC Standard No. 44. There are many recommendations, e.g. U.S. Grade Standards, but they are not mandatory in international trade.

MINIMUM REQUIREMENTS:

Courgettes must be intact, fresh, firm and sound. The stalk of the fruit must be intact, the fruit free from hollows, cracks, developed seeds and without any foreign smell or taste. There must be no signs of attack by pests.

KEEPING QUALITIES:

Shelf life:

10°C, 90% RH, 2-3 weeks
20°C, 60% RH, 3-5 days

Ideal demands: 7-10°C (45-50°F), 90-95% RH

Recommended temperature:

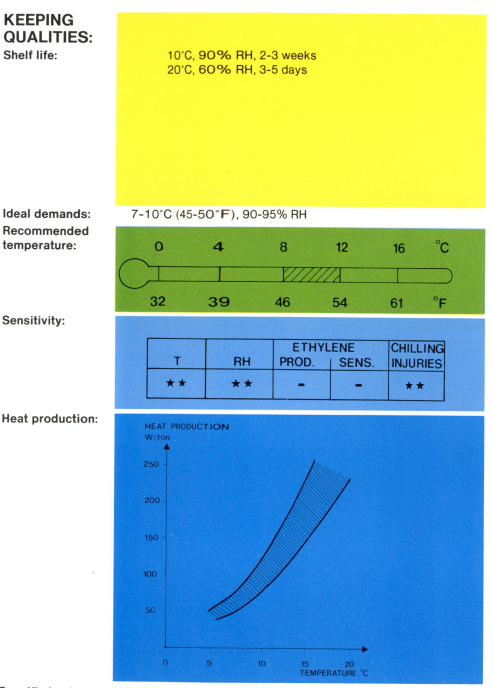

		ETHYLENE		CHILLING
T	RH	PROD.	SENS.	INJURIES
★★	★★	−	−	★★

Heat production:

Specific heat: 3.93 kJ/kg x °C 0.94 Btu/lb x °F

Specific weight: Cartons, pallets 320-400 kg/m³. Bulk approx. 600 kg/m³

CRANBERRY

Latin: Vaccinium macrocarpon Ait.
French: Canneberge or Alrelle d'Amérique
German: Moosbeere or Kranbeere
Spanish: Baya de turbera or Arándano americano

DESCRIPTION OF PRODUCT:

Cranberries comprise 2 varieties both of which originate from North America and northern Asia.

The small–fruited cranberries are most common in Europe and Asia. They look like red whortleberries and are 5-10 mm in diameter, juicy with a sourish taste. They are not normally cultivated, but are gathered in tne wild in damp, acid ground, where they grow wild.

The large cranberries grow principally in North America, where these shrubs are actually cultivated in swampy areas, and the plants irrigated in winter to protect them against frost. The berries are mechanically harvested. The berry of this variety is approx. 2 cm in diameter and has much drier flesh. It is bitter and contains considerable amounts of tannin. Cranberries also contain benzoic acid.

MAJOR PRODUCERS:

The major producers of cranberries are U.S.A., West Germany, Scandinavia and U.S.S.R.

STANDARDS:

There are no international standards for cranberries. Many recommendations are to be found, e.g. U.S. Grade Standards, but they are not mandatory in international trade.

MINIMUM REQUIREMENTS:

Cranberries should be fresh, intact, clean and free from any foreign smell or taste and signs of attack by disease or pests, mould or rot. The berries should be full-bodied and adequately ripe. Berries which are withered, bruised or mechanically damaged should not be included. Leaves, stalks and other inedible parts of the plant should be removed.

Latin: Vaccinium macrocarpon Ait.
French: Canneberge or Alrelle d'Amérique
German: Moosbeere or Kranbeere
Spanish: Baya de turbera or Arándano americano

CRANBERRY

KEEPING QUALITIES:

Shelf life:

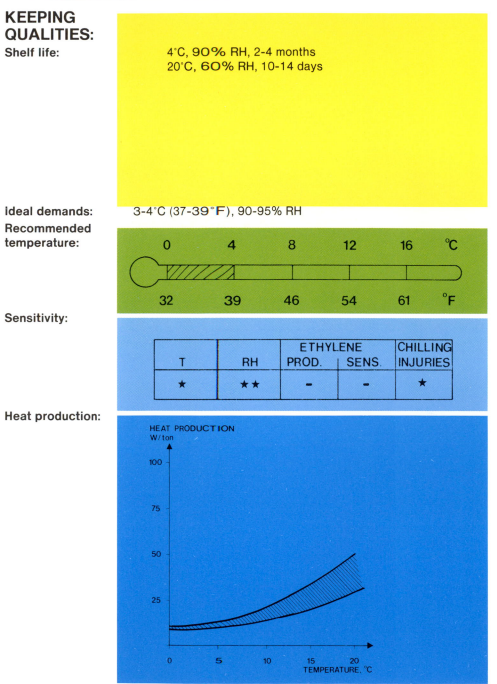

4°C, 90% RH, 2-4 months
20°C, 60% RH, 10-14 days

Ideal demands: 3-4°C (37-39°F), 90-95% RH

Recommended temperature:

0	4	8	12	16	°C
32	39	46	54	61	°F

Sensitivity:

T	RH	ETHYLENE PROD.	SENS.	CHILLING INJURIES
★	★★	−	−	★

Heat production:

HEAT PRODUCTION
W/ton

100

75

50

25

0   5   10   15   20
TEMPERATURE, °C

Specific heat: 3.77 kJ/kg x °C 0.90 Btu/lb x °F
Specific weight: Boxes, pallets 300-350 kg/m³. Bulk approx. 600 kg/m³

115

DESCRIPTION OF PRODUCT

The cucumber plant is an annual vine which has its origin in northern India. The cultivated varieties, which are less bitter than the wild, have been known in China for at least 3000 years.

The cucumis family to which cucumbers belong, consist of countless varieties, and new ones are constantly appearing on the market.

A cucumber is actually an unripe fruit in which the undeveloped seeds lie in 6 rows. Improvements in breeding by producing plants with mostly female flowers that ripen into fruit without fertilization have eliminated bitterness and now only varieties which have an acceptable flavour are produced.

MAJOR PRODUCERS:

The major producers of cucumbers are China, U.S.S.R., Japan, Turkey and Rumania.

STANDARDS:

For trade within the EEC cucumbers must comply with EEC Standard No. 23. Additionally, there are many other recommendations, e.g. U.S. Grade Standards, but they are not mandatory in international trade.

MINIMUM REQUIREMENTS:

Cucumbers should be clean, intact, firm and fresh in appearance and have a green colouring typical of the variety. They should be sufficiently developed but still have soft seeds. The fruit must neither be bitter nor have any foreign smell or taste. Cucumbers should not appear wilted, bruised, discoloured or affected by decay or disease. They should be free from chilling injuries, which may result in sunken blemishes and glassy flesh.

Latin:	Cucumis sativus L.
French:	Concombre
German:	Gurke
Spanish:	Pepino

CUCUMBER

KEEPING QUALITIES:

Shelf life:

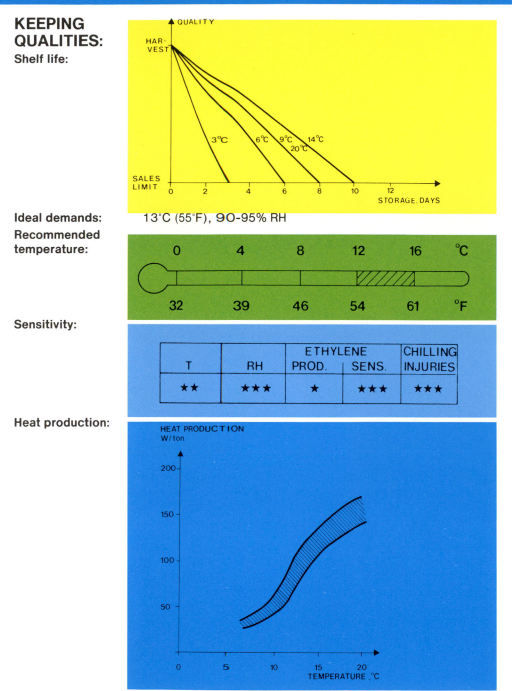

Ideal demands: 13°C (55°F), 90-95% RH

Recommended temperature:

Sensitivity:

T	RH	ETHYLENE PROD.	SENS.	CHILLING INJURIES
★★	★★★	★	★★★	★★★

Heat production:

Specific heat: 4.10 kJ/kg x °C 0.98 Btu/lb x °F

Specific weight: Palletized cartons 320-400 kg/m³. Bulk approx. 600 kg/m³

117

DESCRIPTION OF PRODUCT:

The date is an ancient culture plant, particularly in the African-Asian desert belt. One theory suggests that dates originated in Western India. However, it is certain that dates have been cultivated for 8000 years in India.

The date palm can grow to a height of 20 m. The trunk gets its characteristic scarred look from the constant formation of new leaves while the old fall off leaving large scars. From the large crown of pinnate leaves at the top of the tree hang a number of flower spikes which are wind-pollinated. A fully developed cluster of dates can often weigh around 25 kg.

The fruit of a date is brown, has a firm, smooth skin and a central hard stone. Fresh dates are similar in appearance to dried, but are less sweet.

MAJOR PRODUCERS:

The major producers of dates are Egypt, Saudi Arabia, Iran, Pakistan and Algeria.

STANDARDS:

There are no international standards for fresh dates. Recommendations such as OECD Standards are not mandatory in international trade.

MINIMUM REQUIREMENTS:

Fresh dates should be intact, clean, sound and not affected by disease, insects, rot or mould. The fruit should not be cracked or damaged.

KEEPING QUALITIES:

Shelf life:

0°C, 90% RH, 1-2 months

They may be frozen, but keeping quality will be considerably reduced on thawing.

Ideal demands: 0°C (32°F), 85–90% RH

Recommended temperature:

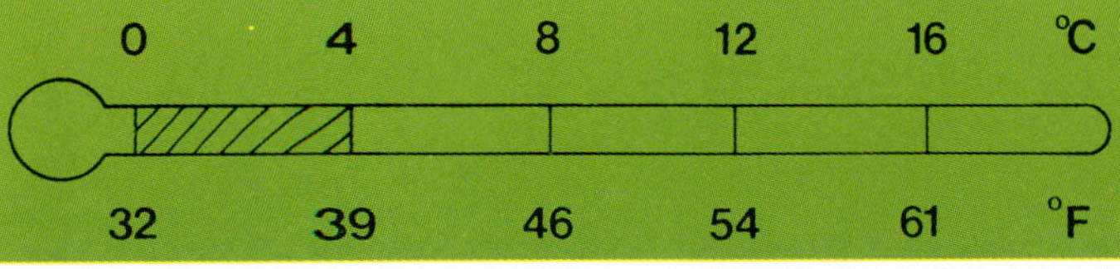

0	4	8	12	16	°C
32	39	46	54	61	°F

Sensitivity:

T	RH	ETHYLENE PROD.	SENS.	CHILLING INJURIES
★ ★	★	-	-	-

Heat production:

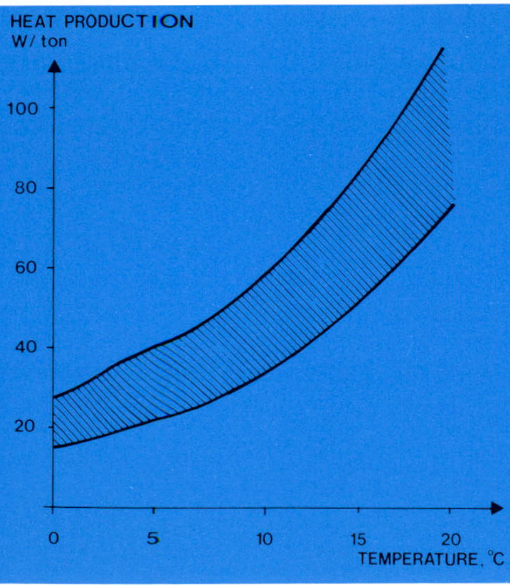

HEAT PRODUCTION
W/ ton

TEMPERATURE, °C

Specific heat: 3.10 kJ/kg x °C 0.74 Btu/lb x °F

Specific weight: Palletized boxes 350-450 kg/m^3 Bulk. approx. 650 kg/m^3

	Latin:	Cichorium endivia L. var. crispum Lam. and
		c. endivia L. var. Latifolium Lam.
ENDIVE and **ESCAROLE**	French:	Chicore frise and chicore scarole
	German:	Krause Endivie and Eskariol
	Spanish:	Escarola repollo or endivia crespa

DESCRIPTION OF PRODUCT:

There are many types of leafy vegetable, often referred to collectively as lettuce, mainly intended for raw consumption. A few of them, of which two are named here, are of significance in international trade.

Endive and escarole originate from wild chicory, still found in southern Europe, Africa and western Asia. Ancient Egyptians used green chicory leaves as vegetables. Its further cultivation has mainly taken place in Europe where it is still widespread.

Escarole is also known as Batavian endive or broad-leafed endive. This type has entire leaves which form a loose head, the heart of which should be yellow.

Endive is also known as curly-leafy endive or frisee. It has more or less divided leaves which form a slightly developed head whose heart should be yellow.

MAJOR PRODUCERS:

The major producers of endive and escarole are France, Italy and Spain, but these products are cultivated all over the world.

STANDARDS:

For trade within the EEC endive and escarole must comply with EEC Standard No. 5. There are also many recommendations, e.g. U.S. Grade Standards, but they are not mandatory in international trade.

MINIMUM REQUIREMENTS:

Endive and escarole should be intact, clean, sound plants which are fresh in appearance. The plants should be normally developed, the root cut directly below the lowest leaf, free from foreign matter and foreign taste or smell. The plant should be well formed with a colour which is typical of the variety, free from pests, disease and frost damage. The heart should be yellow.

Latin: Cichorium endivia L. var. crispum Lam. and
c. endivia L. var. Latifolium Lam.
French: Chicore frise and chicore scarole
German: Krause Endivie and Eskariol
Spanish: Escarola repollo or endivia crespa

ENDIVE and **ESCAROLE**

KEEPING QUALITIES:

Shelf life:

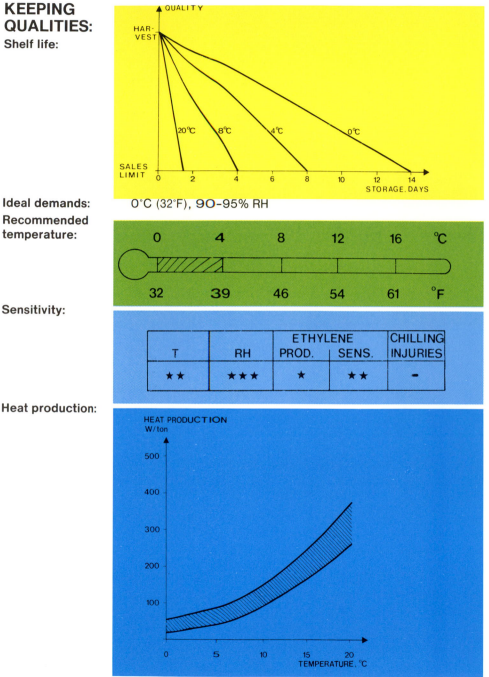

Ideal demands: 0°C (32°F), 90-95% RH

Recommended temperature:

				°C
0	4	8	12	16
32	39	46	54	61 °F

Sensitivity:

T	RH	ETHYLENE PROD.	SENS.	CHILLING INJURIES
★★	★★★	★	★★	-

Heat production:

Specific heat: 4.05 kJ/kg x °C 0.97 Btu/lb x °F
Specific weight: Boxes, pallets 150-200 kg/m³. Bulk approx. 200 kg/m³

121

DESCRIPTION OF PRODUCT:

Feijoa is also called wild guava and has its origin in South America where it still grows wild. From there it was taken to South East Asia and New Zealand where it is commercially cultivated.

Feijoa is a small tree with greyish green leaves. The flowers are red and very decorative. The fruit is an oblong berry differing slightly in shape, according to the variety. The skin is greenish. The flesh nearest to the skin is as firm as that of a peach. In the centre the flesh is jelly-like and contains small seeds which are also edible. It has a sourish taste. A normal fruit is 3-5 cm in diameter.

MAJOR PRODUCERS:

The major producers of feijoa are Brazil, New Zealand, U.S.A. and Israel.

STANDARDS:

There are no international standards for feijoa.

MINIMUM REQUIREMENTS:

Feijoa should be intact, clean and sound.

The fruit should be firm without any signs of mechanical damage or attack by disease or insects. It should have no foreign smell or taste. The colour should be evenly green and the flesh pale and without dark discolourations.

Latin:	Acca sellowiana (Berg) Burret
French:	Feijoa
German:	Fidjoa
Spanish:	Guayaba chilena

FEIJOA (Pineapple guava)

KEEPING QUALITIES:

Shelf life:

10°C, 90% RH, 3 weeks
20°C, 60% RH, 7-10 days

Ideal demands: 8-10°C (46-50°F), 90% RH

Recommended temperature:

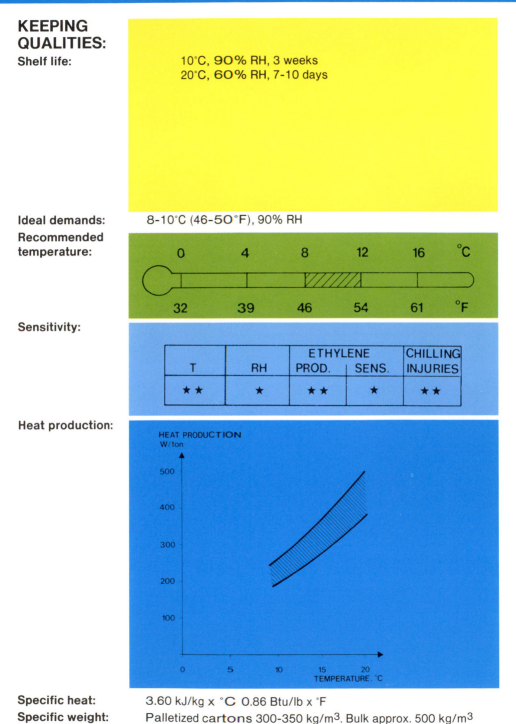

| | 0 | 4 | 8 | 12 | 16 | °C |
| | 32 | 39 | 46 | 54 | 61 | °F |

Sensitivity:

T	RH	ETHYLENE PROD.	SENS.	CHILLING INJURIES
★★	★	★★	★	★★

Heat production:

HEAT PRODUCTION
W/ton

TEMPERATURE. °C

Specific heat: 3.60 kJ/kg x °C 0.86 Btu/lb x °F
Specific weight: Palletized cartons 300-350 kg/m³. Bulk approx. 500 kg/m³

123

(Florence or Italian fennel)
FENNEL

Latin: Foenicullum vulgare Mill. var. dulce
French: Fenouil bulbe
German: Knollenfenchel
Spanish: Hinojo

DESCRIPTION OF PRODUCT:

Fennel has been known as a spice in west Asia and the Mediterranean region since ancient times, but its use as a vegetable is of a much later date.

The whole plant can be consumed. The base of the leaves is thick and forming a characteristic bulb shape with a distinctive anise-like flavour. The finely-divided leaves, normally cut off before sale, can be used as a garnish. The taproot of the fennel is also edible. Fennel is exceptionally rich in vitamin C and carotene. It is normally 6–10 cm in diameter.

MAJOR PRODUCERS:

The major producers of fennel are Italy, France, Spain and Greece.

STANDARDS:

There are no international standards for fennel. However, there are some recommendations, e.g. EEC Standard No. FFV-16, which are not mandatory in international trade.

MINIMUM REQUIREMENTS:

Fennel should be firm, pale and clean with the outer scales covering the whole of its distinctive bulbous form. There should be no discolouration, such as browning of the vascular bundle (strands). The scales should be crisp and without tough fibres or hollowness. The outer scales should be undamaged or so slightly damaged that it does not affect the appearance. The top and the root should be removed with a clean cut.

Latin: Foenicullum vulgare Mill. var. dulce
French: Fenouil bulbe
German: Knollenfenchel
Spanish: Hinojo

(Florence or Italian fennel)
FENNEL

KEEPING QUALITIES:

Shelf life:

0°C, 90% RH, 2-4 weeks
20°C, 60% RH, 2-3 days

Ideal demands:

0°C (32°F), 90–95% RH

Recommended temperature:

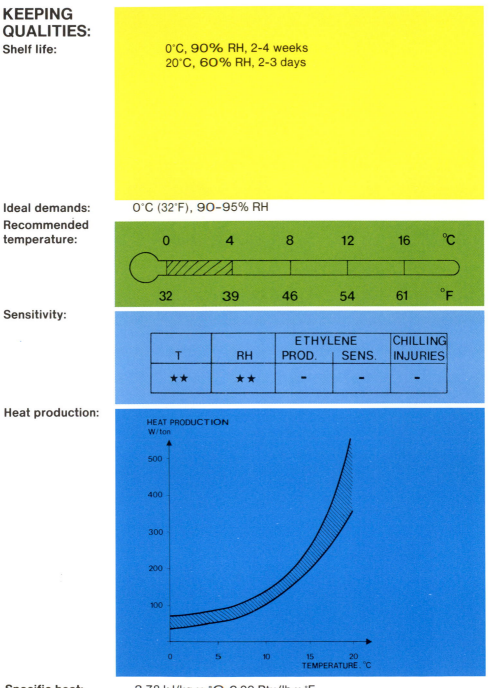

| 0 | 4 | 8 | 12 | 16 | °C |
| 32 | 39 | 46 | 54 | 61 | °F |

Sensitivity:

T	RH	ETHYLENE PROD.	SENS.	CHILLING INJURIES
★★	★★	–	–	–

Heat production:

HEAT PRODUCTION
W/ton

Specific heat: 3.78 kJ/kg x °C 0.90 Btu/lb x °F

Specific weight: Palletized boxes 300-320 kg/m³. Bulk approx. 330 kg/m³

DESCRIPTION OF PRODUCT:

The fig tree has been cultivated in the Mediterranean region since ancient times where figs have played an important dietary role for many thousands of years. They are still common in this region.

It is a deciduous tree which can reach a height of 10 m . The inflorescence is pear-shaped. The floral receptacle forms into a cup around the inside of which are the flowers. The cup is almost closed. Some varieties do not require pollination. Those that do are pollinated by the newly hatched eggs of a small wasp.

The skin of fresh figs is violet and the surface dull. The flesh is pink and contains numerous tiny seeds. Figs with white or green skins that turn yellow during the ripening process are also available, but the violet variety maintains its colour throughout.

Figs are bulb-shaped and have a 4-6 cm diameter. Fresh figs are not as sweet as dried due to high water content.

MAJOR PRODUCERS:

The major producers of figs are Turkey, Italy and the other Mediterranean countries, U.S.A. and Australia.

STANDARDS:

There are no international standards for figs. There are many recommendations, e.g. ECE Standard No. FFV-17, but they are not mandatory in international trade.

MINIMUM REQUIREMENTS:

Figs should be intact, sound, clean, fresh in appearance and free from disease or pests. There should be no foreign smell or taste. The stem may be slightly damaged, but the skin should be unmarked and without cracks or mechanical damage. The surface should be free from spots and show no signs of shrivelling.

KEEPING QUALITIES:

Shelf life

0°C, 90% RH, 1-2 weeks
20°C, 60% RH, 1-2 days

Ideal demands: 0°C (32°F), 90-95% RH

Recommended temperature:

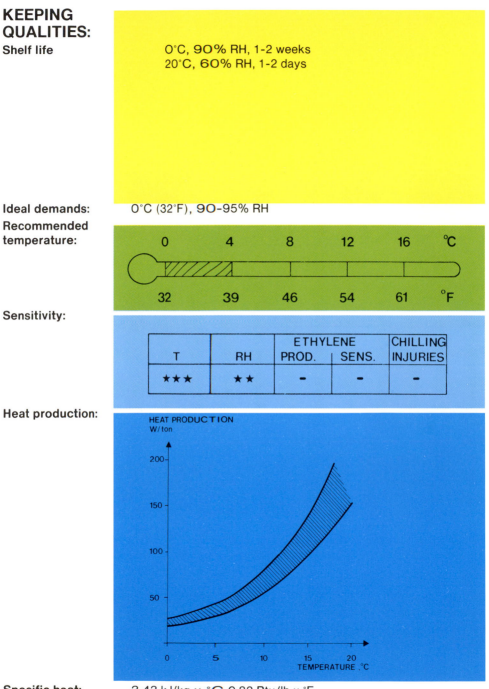

| 0 | 4 | 8 | 12 | 16 | °C |

| 32 | 39 | 46 | 54 | 61 | °F |

Sensitivity:

T	RH	ETHYLENE PROD.	SENS.	CHILLING INJURIES
★★★	★★	-	-	-

Heat production:

HEAT PRODUCTION
W/ton

200 –

150 –

100 –

50 –

0   5   10   15   20
TEMPERATURE ‚°C

Specific heat: 3.43 kJ/kg x °C 0.82 Btu/lb x °F
Specific weight: Palletized cartons 150-200 kg/m³

127

GARLIC

Latin:	Allium sativum L.
French:	Ail
German:	Knoblauch
Spanish:	Ajo

DESCRIPTION OF PRODUCT:

Garlic is not found growing wild, but it is believed to have its origin in Central Asia. Garlic was widespread in the Mediterranean region at a very early stage. Cultivation of garlic is known to have existed 5000 years. The regions where garlic is cultivated today lie especially in the subtropical belt, but there is also considerable production in the tropical and temperate regions.

Garlic seldom produces seeds and is propagated by planting the cloves of which the garlic bulb is made up. Unlike the round leaves of the onion, the garlic plant has flat leaves and can be compared with those of the leek.

Garlic has a strong, distinctive odour and taste which is attributable to its sulphur-containing essential oils.

MAJOR PRODUCERS:

The major producers of garlic are China, South Korea, Spain, India and Thailand.

STANDARDS:

For trade within the EEC garlic must comply with EEC Standard No. 28. There are several other recommendations, e.g. U.S. Grade Standards, but they are not mandatory in international trade.

MINIMUM REQUIREMENTS:

Garlic should be sound, clean and firm. The bulbs should be free from frost damage, sunburn, disease, rot and mould. No sprouts should be visible. Garlic must be dried thoroughly, but without the flesh losing its firmness. The bulbs should be intact with no cloves missing. They should be free of any foreign smell or taste.

KEEPING QUALITIES:

Shelf life:

0°C, **70%** RH, 6-7 months
20°C, **60%** RH, 3-4 weeks

Ideal demands: 0°C (32°F), **65–70%** RH

Recommended temperature:

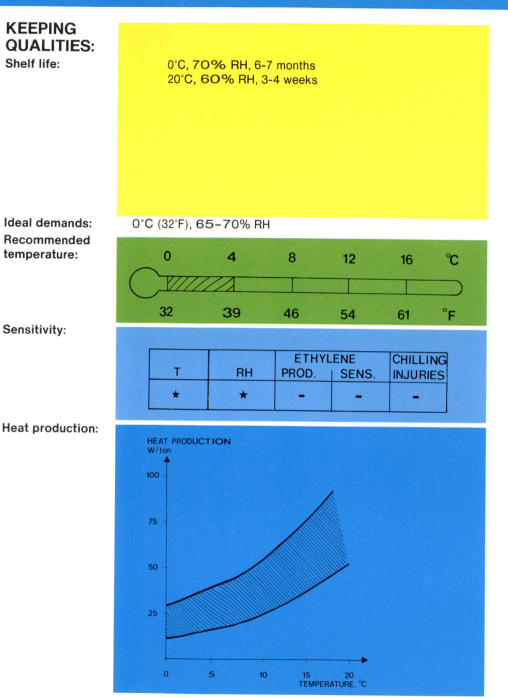

| | 0 | 4 | 8 | 12 | 16 | °C |
| 32 | 39 | 46 | 54 | 61 | °F |

Sensitivity:

T	RH	ETHYLENE PROD.	SENS.	CHILLING INJURIES
★	★	–	–	–

Heat production:

HEAT PRODUCTION
W/ton

TEMPERATURE, °C

Specific heat: 2.89 kJ/kg x °C 0.69 Btu/lb x °F
Specific weight: Nets, pallets **320**-400 kg/m³ Bulk approx. 550 kg/m³

DESCRIPTION OF PRODUCT:

Grapes originally came from south-west Asia - Armenia - via Asia Minor, to the Mediterranean countries and to the rest of the world.

Grapes are normally cultivated as bushes. The grapes hang in close clusters which generally weigh 100-1000 g. Each grape is covered with bloom. This waxy layer protects the grapes from withering. The fruits usually contain a few seeds, but special seedless varieties for the fresh market have been developed.

MAJOR PRODUCERS:

The major producers of grapes are Italy, France, U.S.S.R., Spain and U.S.A.

STANDARDS:

For trade within the EEC grapes must comply with EEC Standard No. 14. There are, however, many recommendations, e.g. U.S. Grade Standards, but they are not mandatory in international trade.

MINIMUM REQUIREMENTS:

Grapes should be sound, clean and in-
tact. Each grape should hang firmly in a cluster. The grapes should be undamaged, free from any signs of attack by disease and insects and from visible remnants of chemical agents. They should not emit any foreign smell or taste. Shelf life can be extended by fumigation of the grapes with sulphur dioxide. This procedure is, however, illegal in many countries.

Latin: Vitis vinifera L.
French: Raisin
German: Weintraube or Tafeltraube
Spanish: Uva (de mesa)

KEEPING QUALITIES:

Shelf life:

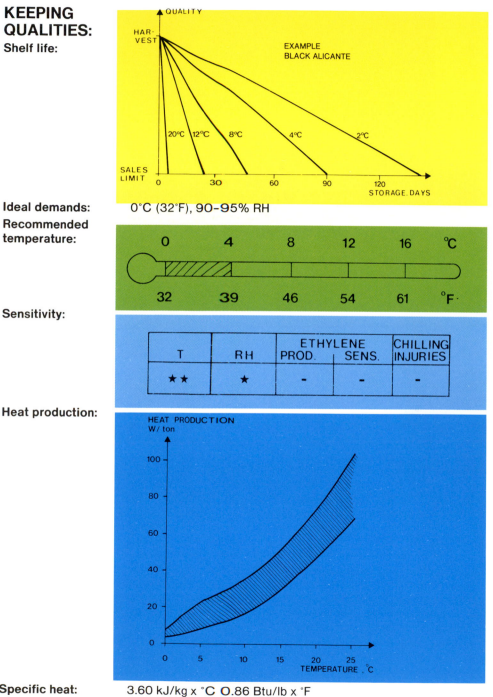

QUALITY

HAR-
VEST

EXAMPLE
BLACK ALICANTE

20°C 12°C 8°C 4°C 2°C

SALES
LIMIT

0   30   60   90   120

STORAGE. DAYS

Ideal demands: 0°C (32°F), 90–95% RH

Recommended temperature:

| 0 | 4 | 8 | 12 | 16 | °C |
| 32 | 39 | 46 | 54 | 61 | °F |

Sensitivity:

T	RH	ETHYLENE PROD.	SENS.	CHILLING INJURIES
★★	★	–	–	–

Heat production:

HEAT PRODUCTION
W/ ton

100

80

60

40

20

0

0   5   10   15   20   25

TEMPERATURE , °C

Specific heat: 3.60 kJ/kg x °C 0.86 Btu/lb x °F

Specific weight: Palletized boxes 330–400 kg/m^3. Bulk, cluster approx. 300 kg/m^3

GRAPEFRUIT

Latin:	Citrus x paradisi Macfad.
French:	Pamplemousse
German:	Grapefruit or Pampelmuse
Spanish:	Pomelo

DESCRIPTION OF PRODUCT:

Grapefruit is presumed to be a hybrid of orange and pomelo (pampelmus). The pomelo probably came from Indonesia, and spread to South East Asia and Oceania. In the 17th century Captain Shaddock brought the pomelo to the West Indies where grapefruit is said to have come into existence. Grapefruit was first described in the 18th century in Puerto Rico.

Almost all the known varieties come from Florida. Now grapefruit is grown almost all over the world in subtropical and tropical regions.

The grapefruit tree can be 5 m tall and can bear several hundred kilogrammes of fruit each year. The fruits grow in clusters like grapes, hence the name, grapefruit. Earlier grapefruit contained numerous seeds, but hybridization has produced varieties that are almost free of seeds. The fruit contains a small amount of quinine which gives it its bitter taste.

MAJOR PRODUCERS:

The major producers of grapefruit are U.S.A., Israel, Argentina, Cuba and Cyprus.

STANDARDS:

There are no international standards for grapefruit. There are several recommendations, e.g. U.S. Grade Standards, but they are not mandatory in international trade.

MINIMUM REQUIREMENTS:

Grapefruit must be intact, sound, free of frost and other damage. It must be free of any visible foreign matter and of any foreign smell or taste. It should be without blemishes or discolouration of the rind or of the flesh, which would reduce the fruit suitability for consumption. Grapefruit should bear no signs of chilling damage, which can also result in sunken spots or brown discolouration of the flesh.

Latin:	Citrus x paradisi Macfad.
French:	Pamplemousse
German:	Grapefruit or Pampelmuse
Spanish:	Pomelo

KEEPING QUALITIES:

Shelf life:

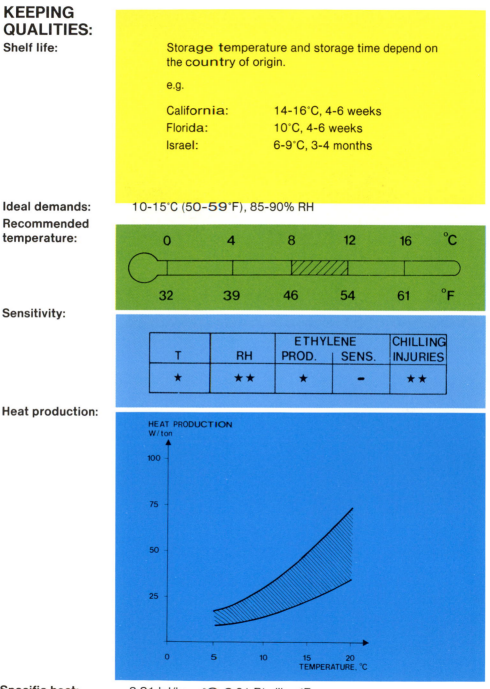

Storage temperature and storage time depend on the country of origin.

e.g.

California:	14-16°C, 4-6 weeks
Florida:	10°C, 4-6 weeks
Israel:	6-9°C, 3-4 months

Ideal demands: 10-15°C (50-59°F), 85-90% RH

Recommended temperature:

0	4	8	12	16	°C
32	39	46	54	61	°F

Sensitivity:

T	RH	ETHYLENE PROD.	SENS.	CHILLING INJURIES
★	★★	★	–	★★

Heat production:

HEAT PRODUCTION W/ton

(graph: vertical axis 25, 50, 75, 100; horizontal axis TEMPERATURE, °C: 0, 5, 10, 15, 20)

Specific heat: 3.81 kJ/kg x °C 0.91 Btu/lb x °F

Specific weight: Palletized boxes 350-450 kg/m^3. Bulk approx. 500 kg/m^3

Latin:	Psidium guajava L.
French:	Goyave
German:	Guave or Guajave
Spanish:	Guayaba

DESCRIPTION OF PRODUCT:

The guava is native to the tropical region of South America where it is still found growing wild. It is now grown all over the tropical regions.

Guava is easy to cultivate. The seeds are spread by birds, and to such an extent that in some areas they grow like weeds.

The guava tree is an evergreen which can reach a height of 10 m. The fruit comes in several colours and shapes, but they are normally round or oval with a diameter of about 6 cm.

The rind is rough and cannot be eaten. The pulp is white or slightly orange-red. There are several hundred seeds in the middle of the fruit. When it is ripe it yields to slight pressure.

MAJOR PRODUCERS:

The major producers of guava are Brazil, U.S.A., Israel and India.

STANDARDS:

There are no international standards for guava.

MINIMUM REQUIREMENTS:

Guavas should be intact, sound and clean. They should not be infected by disease, insects, rot or mould. The fruit should be full-bodied and bear no signs of damage or pressure marks.

Latin:	Psidium guajava L.
French:	Goyave
German:	Guave or Guajave
Spanish:	Guayaba

GUAVA

KEEPING QUALITIES:

Shelf life:

10°C, 90% RH, 3 weeks
20°C, 60% RH, 7-10 days

Ideal demands: 8-10°C (46-50°F), 90% RH

Recommended temperature:

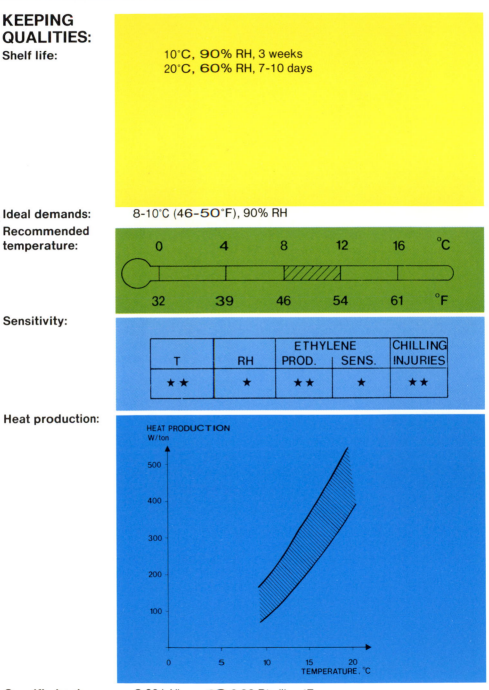

Sensitivity:

T	RH	ETHYLENE PROD.	SENS.	CHILLING INJURIES
★★	★	★★	★	★★

Heat production:

Specific heat: 3.60 kJ/kg x °C 0.86 Btu/lb x °F

Specific weight: Palletized cartons 300-350 kg/m³. Bulk approx. 500 kg/m³

HORSE RADISH

Latin:	Armoracia rusticana Ph. Gartn., B. Mey et Scherb
French:	Raifort
German:	Meerrettich
Spanish:	Rábano silvestre

DESCRIPTION OF PRODUCT:

The origin of horse radish is unknown. It is presumed to have come from Asia and brought to Europe approx. 1000 years ago.

The horse radish plant is a herb with one-meter long leaves. Botanically it is a perennial, but industrially cultivated as an annual. The approx. 1 cm thick and 30 cm long roots, which are branch roots from the previous year's rootstock, are planted in the spring. Horse radish is harvested in September-November and consists of a long main root, with several long and thin branch roots which may eventually be used as next year's crop.

Horse radish is rich in Vitamin C. The pungent taste is due to the presence of the glucoside, sinigrin. This substance is split up into allyl mustard oil with the help of enzymes.

MAJOR PRODUCERS:

The major producers of horse radish are Poland, Hungary and U.S.S.R.

STANDARDS:

There are no international standards for horse radish. There are many recommendations, e.g. ECE Standard No. FFV-20 and U.S. Grade Standards, but they are not mandatory in international trade.

MINIMUM REQUIREMENTS:

Horse radish should only comprise pieces of sturdy root without any above-ground parts. They should be firm and full-bodied. The root parts should be clean and without branching or woody pieces. There should be no signs of rust or other forms of discolouration. Horse radish should be sound, free from attack by disease, pests, rot, mould, damage and foreign smell or taste.

Latin:	Armoracia rusticana Ph. Gartn., B. Mey et Scherb
French:	Raifort
German:	Meerrettich
Spanish:	Rábano silvestre

HORSE RADISH

KEEPING QUALITIES:

Shelf life:

-1-0°C, 95% RH, 10-12 months
20°C, 60% RH, 7-10 days

Ideal demands: -1-0°C (30-32°F), 90-95% RH

Recommended temperature:

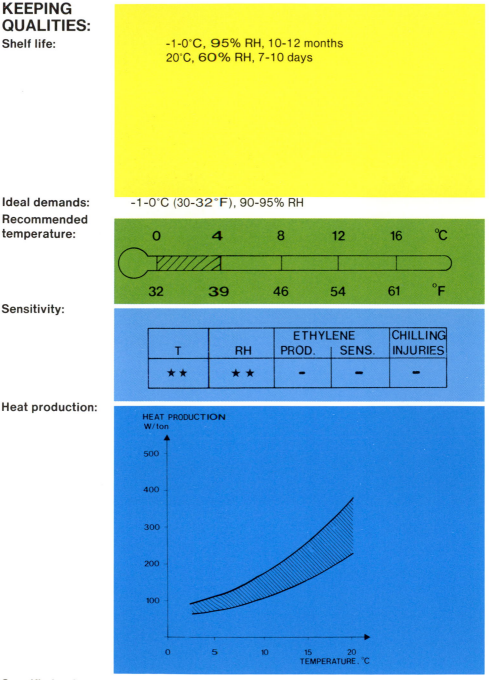

Sensitivity:

| | | ETHYLENE | | CHILLING |
T	RH	PROD.	SENS.	INJURIES
★★	★★	-	-	-

Heat production:

Specific heat: 3.35 kJ/kg x °C   0.80 Btu/lb x °F

Specific weight: Boxes, pallets 250-350 kg/m³. Bulk approx. 500 kg/m³

JACKFRUIT

Latin: Artocarpus heterophyllus Lam.
French: Jacquier
German: Jackfrucht
Spanish: Artocarpa or Jaqueira

DESCRIPTION OF PRODUCT:

Jackfruit has its origin in India from where it spread to South East Asia and the Philippines. Today it is widespread in the tropics.

The jackfruit tree, which has big, decorative leaves, is an evergreen and can reach a height of 20 m. Its fruits are enormous and may weigh up to 50 kg, but most varieties bear fruit that weigh 2-10 kg. Fruits can be harvested three times a year from the same tree. The tree provides wood and the young bark raffia for weaving.

The jackfruit is a large, oblong or irregularly shaped fruit. The skin is thick and has a bluntly spiked surface. The edible flesh lies in pigeon-egg sized formations (embedded in an inedible matrix) each containing a large seed, around a central stem. The flesh is yellowish or slightly lilac. A ripe jackfruit is soft and fleshy and has a sweet-sour and very characteristic taste.

MAJOR PRODUCERS:

The major producers of jackfruits are India, Sri Lanka, Indonesia, Malaysia and the Philippines.

STANDARDS:

There are no international standards for jackfruit.

MINIMUM REQUIREMENTS:

Jackfruit should be intact and sound and its skin undamaged and crack-free. Attack by disease or insects, rot or mould should not occur. The flesh should be fair, without discolouration or any brown parts.

138

KEEPING QUALITIES:

Shelf life:

13°C, 90% RH, 1-3 weeks
20°C, 60% RH, 3-5 days

Ideal demands: 13°C (55°F), 90-95% RH

Recommended temperature:

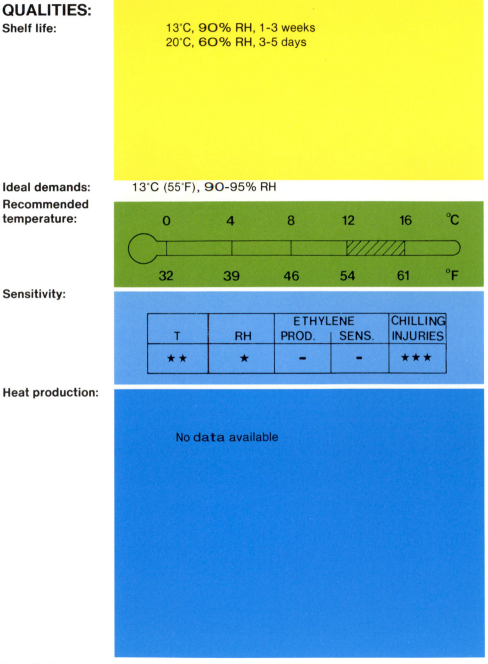

| | 0 | 4 | 8 | 12 | 16 | °C |
| | 32 | 39 | 46 | 54 | 61 | °F |

Sensitivity:

T	RH	ETHYLENE PROD.	SENS.	CHILLING INJURIES
★★	★	−	−	★★★

Heat production:

No data available

Specific heat: 3.26 kJ/kg x °C 0.78 Btu/lb x °F

Specific weight: Palletized boxes 200-300 kg/m^3. Bulk approx. 550 kg/m^3

ARTICHOKE, JERUSALEM

Latin:	Helianthus tuberosus L.
French:	Topinambour
German:	Erdbirne or Topinambur
Spanish:	Aguaturma

DESCRIPTION OF PRODUCT:
Jerusalem artichokes originate from North America and came to Europe via Asia in the 17th century.

The plants is a perennial, can grow to a height of 2 m and bears egg or heart-shaped leaves. The flowers, which are relatively small, yellow and resemble the sunflower, bloom only under very warm summer conditions. The tuberous, edible roots form at the tip of short subterranean runners. The tubers do not contain starch as potatoes do, but insulin composed of units of fructose. This is of significance to diabetics.

MAJOR PRODUCERS:
The major producers of Jerusalem artichokes are U.S.A., France, West Germany and the Netherlands.

STANDARDS:
There are no international standards for Jerusalem artichokes.

MINIMUM REQUIREMENTS:
Jerusalem artichokes should be well rounded, fresh in appearance, sound and without foreign smell or taste. The tubers should be relatively regularly shaped and should not be fibrous or woody. They should be full-bodied and show no signs of mechanical damage or attack by disease or pests.

KEEPING QUALITIES:

Shelf life:

0°C, **90%** RH, 3-5 months
20°C, **60%** RH, 1-2 weeks

Ideal demands: 0°C (32°F), 90–95% RH

Recommended temperature:

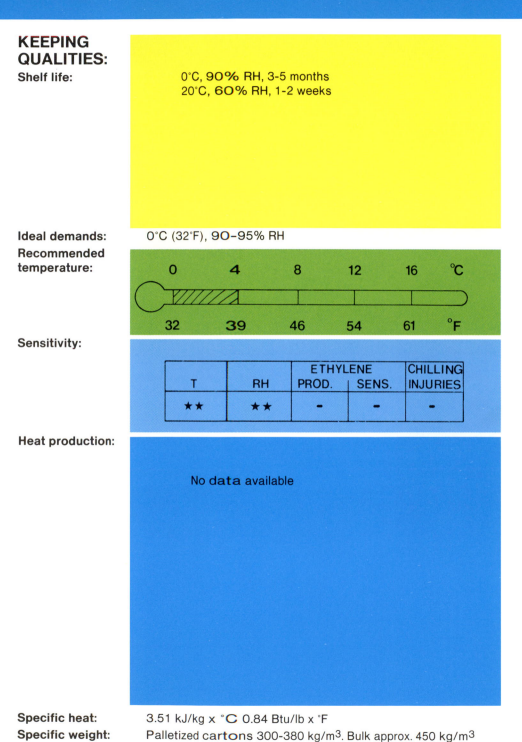

Sensitivity:

T	RH	ETHYLENE PROD.	SENS.	CHILLING INJURIES
★★	★★	-	-	-

Heat production:

No data available

Specific heat: 3.51 kJ/kg x °C 0.84 Btu/lb x °F

Specific weight: Palletized cartons 300-380 kg/m³. Bulk approx. 450 kg/m³

KAKI (Persimmon)

Latin:	Diospyros kaki L.
French:	Kaki or Figue Caque
German:	Kakipflaume or Dattelpflaume
Spanish:	Caqui or Kaki

DESCRIPTION OF PRODUCT:

The kaki tree is deciduous and has its origin in eastern Asia, probably China or Japan. The name, kaki, means fruit in Japanese. Kaki was brought to southern France from where its cultivation spread to the other Mediterranean countries and California. Today kaki is grown almost all over the world where the climate is suitably mild.

The kaki fruit resembles a tomato, but is identifiable by its four large sepals. The colour of the skin and pulp can be yellow or reddish orange. The fruit has up to 8 chambers.

When the fruit is fully ripened the pulp is glassy and of a very soft, almost liquid consistency. The taste is sweet. Most varieties are seed-free and, when unripe, contain a bitter compound which is broken down during the ripening process. Therefore, these varieties are not edible until the fruit is soft and ripe.

There are varieties which contain no bitter compounds and these can be eaten while the fruit is still firm. Some other fruits are stored in a high carbon dioxide atmosphere for 24 hours to remove the bitter compounds before sales. Dark spots sometimes form in the flesh during this treatment. This does not affect the taste in any way, but only the appearance.

MAJOR PRODUCERS:

The major producers of kaki are Japan, China, U.S.A., Israel and Brazil.

STANDARDS:

There are no international standards for kaki.

MINIMUM REQUIREMENTS:

Kaki should be intact, clean, fresh and without foreign smell or taste. The stalk and the sepals should be intact, the skin undamaged and unblemished without bruises or pressure marks. The fruit should not show signs of attack by disease or pests and the flesh should be free of black spots.

Latin: Diospyros kaki L.
French: Kaki or Figue Caque
German: Kakipflaume or Dattelpflaume
Spanish: Caqui or Kaki

(Persimmon) **KAKI**

KEEPING QUALITIES:

Shelf life:

0°C, **90%** RH, 2-3 months
20°C, **60%** RH, 5-6 days

Ideal demands: 0°C (32°F), **90**–95% RH

Recommended temperature:

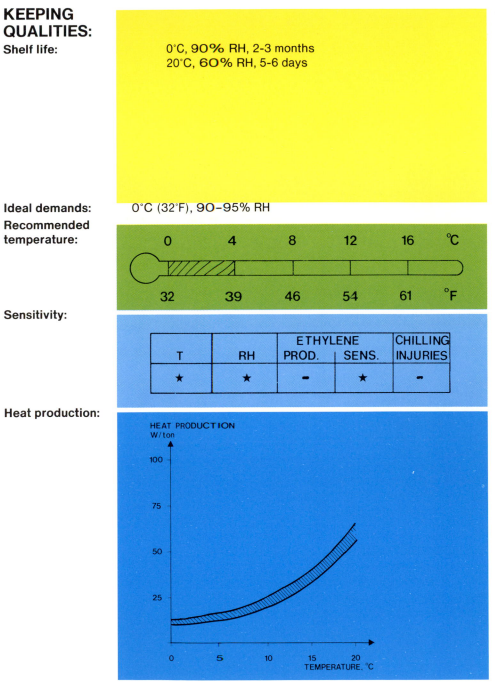

	T	RH	ETHYLENE PROD.	SENS.	CHILLING INJURIES
	★	★	–	★	–

Sensitivity:

Heat production:

HEAT PRODUCTION
W/ton

Specific heat: 3.47 kJ/kg x °C 0.83 Btu/lb x °F
Specific weight: Boxes, pallets 300-350 kg/m³. Bulk approx. 550 kg/m³

DESCRIPTION OF PRODUCT:

The kiwi came originally from China. The plant is a vine that is mainly cultivated in subtropical regions.

The fruit is round, slightly oval, around 5 cm in diameter. It is in fact a berry with a thin, brown and hairy skin. The flesh is green and juicy with a number of small, dark seeds.

The fruit is picked unripe. The unripe fruit may also be eaten, but it is slightly sour. A kiwi fruit is ripe when it yields to a gentle pressure.

MAJOR PRODUCERS:

The most important producer of the kiwi fruit is New Zealand, but U.S.A., France and Israel are now also major producers.

STANDARDS:

There are no international standards for kiwi fruit. However, there are several recommendations, e.g. U.S. Grade Standards, but they are not mandatory in international trade.

MINIMUM REQUIREMENTS:

Kiwi should be intact, clean and sound. The fruit should be firm and without any cracks, pressure marks or signs of mechanical damage. Kiwi should not be infected by rot or mould, nor damaged by frost causing mushy flesh, nor by ethylene causing soft, abnormally pale flesh.

KEEPING QUALITIES:

Shelf life:

0°C, 2-3 months
20°C, 7-10 days

Ideal demands: 0°C (32°F), 90-95% RH

Recommended temperature:

0	4	8	12	16	°C
32	39	46	54	61	°F

Sensitivity:

T	RH	ETHYLENE PROD.	SENS.	CHILLING INJURIES
★	★	★	★★	-

Heat production:

HEAT PRODUCTION
W/ ton

Specific heat: 3.64 kJ/kg x °C 0.87 Btu/lb x °F
Specific weight: Boxes, pallets 300-350 kg/m³. Bulk approx. 550 kg/m³

KUMQUAT

Latin: Fortunella japonica (Thunb.) Swingle
French: Kumquat
German: Kumquat or Zwergorange
Spanish: Kumquat

DESCRIPTION OF PRODUCT:

Kumquat is native to China where it is still found growing wild. From here it spread to Japan, the Mediterranean region and Brazil. In some places it has won popularity as an indoor plant due to its size and decorativeness.

Kumquat is a small tree (approx. 1.5 m in height); the small citrus fruits are 2-4 cm long. The orange-coloured rind is quite thin and contains the juicy pulp and a number of seeds. Its segmented flesh resembles other citrus fruits. The fruit is eaten with its rind. Kumquat tastes spicy and sourish.

There are 2 types of kumquat, round and oval.

MAJOR PRODUCERS:

The major producers of kumquat are Brazil, Israel, Italy and Central America.

STANDARDS:

There are no international standards for kumquat.

MINIMUM REQUIREMENTS:

Kumquat should be intact, clean and sound, its colour an even yellowish orange. No signs of damage, attack by disease or insects should be visible. The fruit should be free from foreign smell or taste.

Latin: Fortunella japonica (Thunb.) Swingle
French: Kumquat
German: Kumquat or Zwergorange
Spanish: Kumquat

KUMQUAT

KEEPING QUALITIES:

Shelf life:

10°C, 90% RH, 4 weeks
20°C, 60% RH, 1 week

Ideal demands:

10°C (50°F), 90-95% RH

Recommended temperature:

0	4	8	12	16	°C
32	39	46	54	61	°F

Sensitivity:

T	RH	ETHYLENE PROD.	SENS.	CHILLING INJURIES
★★	★★	★	–	★★

Heat production:

HEAT PRODUCTION
W/ ton

TEMPERATURE, °C

Specific heat: 3.64 kJ/kg x °C 0.87 Btu/lb x °F
Specific weight: Boxes, pallets 300-350 kg/m³. Bulk 450-550 kg/m³

DESCRIPTION OF PRODUCT:

Leek originated in the Mediterranean countries where it has been cultivated for thousands of years.

Leek is a biennial. The long leaves form a sheath or shaft at the lower end of the plant which is the valuable part.

In its second year , a large, tough and woody stalk grows from the centre of the plant rendering the leek inedible.

MAJOR PRODUCERS:

The major producers of leek are France, the Netherlands, Turkey and Spain.

STANDARDS:

For trade within the EEC leek must comply with EEC Standard No. 39. There are many recommendations, but they are not mandatory in international trade.

MINIMUM REQUIREMENTS:

Leek should be whole, fresh and sound without shoots. Leek should be free from soil, rot, pests and foreign matter and without foreign smell or taste. The leaves should be firm and free from damage and disease.

KEEPING QUALITIES:

Shelf life:

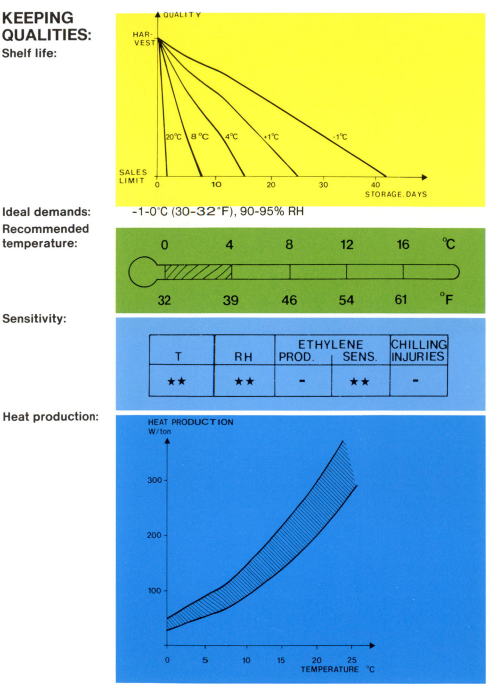

Ideal demands: -1-0°C (30–32°F), 90-95% RH

Recommended temperature:

0	4	8	12	16	°C
32	39	46	54	61	°F

Sensitivity:

T	RH	ETHYLENE PROD.	SENS.	CHILLING INJURIES
★★	★★	–	★★	–

Heat production:

Specific heat: 3.93 kJ/kg x °C 0.94 Btu/lb x °F

Specific weight: Palletized boxes 200-240 kg/m3. Palletized sacks 260-300 kg/m3. Bulk approx. 400 kg/m3

LEMON

Latin: Citrus limon (L.) Burm. f.
French: Citron
German: Zitrone
Spanish: Limón

DESCRIPTION OF PRODUCT:

The lemon **tree** is native to northern India and has been cultivated in Asia Minor and China for several thousand years. The lemon **tree** blossoms all year round and, therefore, blossoms and sets fruit simultaneously. During summer the tree does not normally set fruit, but by means of a special cultivation technique fruiting is delayed in such a manner that lemons can be harvested the rest of the year. Ripe lemons are yellow. They are egg-shaped with a 50-80 mm diameter. The rind is leathery and is often 5-10 mm thick.

MAJOR PRODUCERS:

The major producers of lemons are Italy, U.S.A. and Spain.

STANDARDS:

In trading within the EEC lemons must comply with EEC Standard No. 18. There are many additional recommendations, e.g. U.S. Grade Standards, but they are not mandatory in international trade.

MINIMUM REQUIREMENTS:

Lemons should be whole, sound and free from damage by frost. No foreign matter should be visible and there should be no trace of any foreign smell or taste. The juice content should be min. 25% of the fruit's total weight. Lemons should bear no signs of chilling injury which can result in brown and dry flesh.

150

KEEPING QUALITIES:

Shelf life:

Green:	14-15°C, 1-4 months
Yellow:	11-13°C, 3-6 weeks
	20°C, 1-3 weeks

Ideal demands: 11-15°C (52-59°F), 85-90% RH Green over 14°C (57°F)

Recommended temperature:

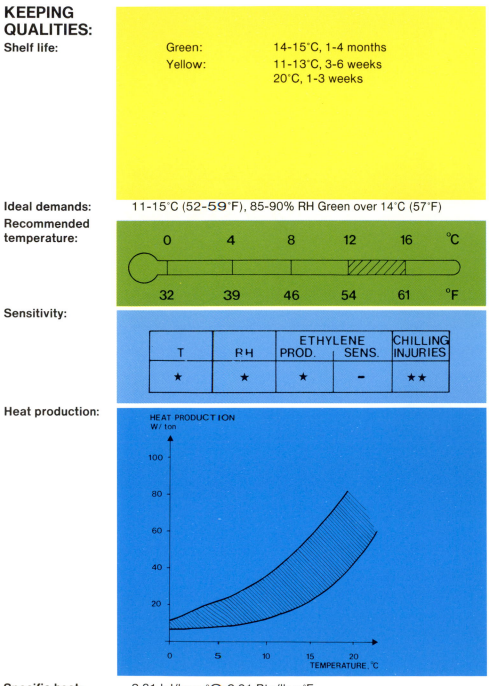

T	RH	ETHYLENE PROD.	SENS.	CHILLING INJURIES
★	★	★	–	★★

Heat production:

HEAT PRODUCTION W/ton

TEMPERATURE, °C

Specific heat: 3.81 kJ/kg x °C 0.91 Btu/lb x °F

Specific weight: Palletized cartons 350-450 kg/m³. Bulk 480-520 kg/m³

(Cabbage lettuce)

LETTUCE, BUTTERHEAD

Latin: Lactuca sativa L. var. capitata L.
French: Laitue pomme
German: Kopfsalat
Spanish: Lechuga repolluda

DESCRIPTION OF PRODUCT:

Lettuce probably originates from the wild varieties that are found in southern Europe and in the Mediterranean region.

There is a large number of varieties that are adapted to various light and temperature conditions. The leaves grow in the form of a rosette and, as the head develops, it gradually becomes firm and compact. The compact heads have light yellow or almost colourless inner leaves. A normal head weighs approx. 150 g.

MAJOR PRODUCERS:

The major producers of lettuce are Italy, France, Spain, U.K. and the Netherlands.

STANDARDS:

For trade within the EEC head lettuce must comply with EEC Standard No. 5. There are many recommendations, e.g. U.S. Grade Standards, but they are not mandatory in international trade.

MINIMUM REQUIREMENTS:

Lettuce must be crisp, firm and look fresh. The heads should be free from damage, insects and disease. Lettuce should be green. The head should be cut off directly below the lowest leaf.

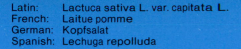

Latin: Lactuca sativa L. var. capitata L.
French: Laitue pomme
German: Kopfsalat
Spanish: Lechuga repolluda

(Cabbage lettuce)
BUTTERHEAD **LETTUCE**

KEEPING QUALITIES:

Shelf life:

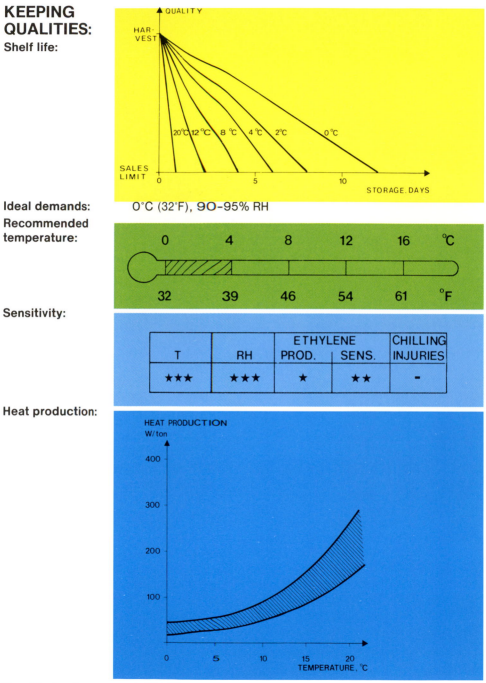

Ideal demands: 0°C (32°F), 90–95% RH

Recommended temperature:

0	4	8	12	16	°C
32	39	46	54	61	°F

Sensitivity:

T	RH	ETHYLENE PROD.	SENS.	CHILLING INJURIES
★★★	★★★	★	★★	-

Heat production:

Specific heat: 4.08 kJ/kg x °C 0.98 Btu/lb x °F

Specific weight: Palletized cartons 120-140 kg/m³. Bulk approx. 250 kg/m³

(Crisphead lettuce)

LETTUCE, ICEBERG

Latin:	Lactuca sativa L. var. capitata
French:	Laitue pommée frisée
German:	Eisbergsalat or Eissalat
Spanish:	Lechuga crespa de hielo

DESCRIPTION OF PRODUCT:

Iceberg lettuce is believed to originate from the wild lettuce that can be found in southern Europe and the Mediterranean region. As an independent variety it is thought to have come from France where it has been known for approx. 200 years.

There are numerous varieties of this lettuce that have been adapted to various light and temperature conditions. The leaves, which are thick and crispy, grow closely to form a rosette that gradually hearts-up. In mature heads the inner leaves are light yellow or almost colourless. A normal head weighs around 300 g.

MAJOR PRODUCERS:

The major producers of iceberg lettuce are U.S.A., Spain, Israel, France and the Netherlands.

STANDARDS:

For trade within the EEC iceberg lettuce must comply with EEC Standard No. 5. There are several recommendations, e.g. U.S. Grade Standards, but they are not mandatory in international trade.

MINIMUM REQUIREMENTS:

Iceberg lettuce heads should be crisp, firm and fresh in appearance. They should be green and without discolouration of the outer leaves or the inner. The heads should be free from damage, pests and disease. They should be cut off just below the basal leaves.

Latin: Lactuca sativa L. var. capitata
French: Laitue pommée frisée
German: Eisbergsalat or Eissalat
Spanish: Lechuga crespa de hielo

(Crisphead lettuce)
ICEBERG **LETTUCE**

**KEEPING
QUALITIES:**
Shelf life:

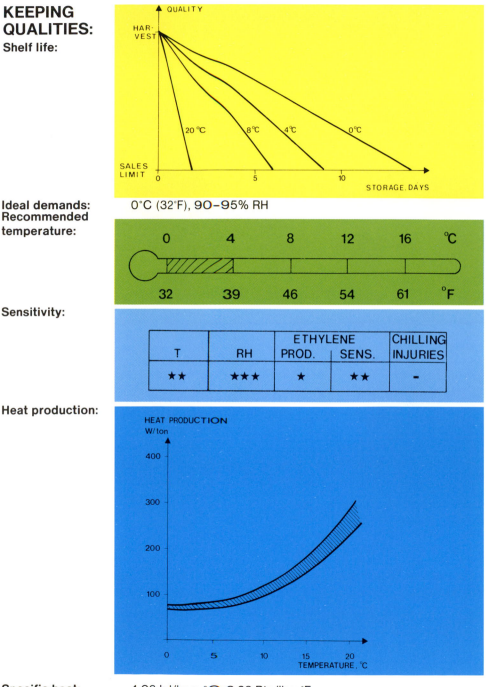

**Ideal demands:
Recommended
temperature:**

0°C (32°F), 90–95% RH

Sensitivity:

| | | ETHYLENE | | CHILLING |
T	RH	PROD.	SENS.	INJURIES
★★	★★★	★	★★	–

Heat production:

Specific heat: 4.08 kJ/kg x °C 0.98 Btu/lb x °F
Specific weight: Palletized boxes 150-200 kg/m³. Bulk approx. 250 kg/m³

155

LIME

Latin:	Citrus aurantiifolia (Christm.) Swingle
French:	Lime or Limette
German:	Limette or Limone
Spanish:	Lima or Limón dulce

DESCRIPTION OF PRODUCT:

The lime fruit has its origin in South East Asia, probably Malaysia. From there it spread to India and then to Europe and America, where it grows wild in many areas. Lime is now cultivated especially in Central America and certain parts of Asia.

The lime tree is small and very spiny. The evergreen leaves are small and light green.

The lime fruit, both the rind and its juicy flesh, is refreshingly green. Limes are oval, structured in segments like oranges, taste sourish and are more aromatic than lemons. They are approx. 5 cm in diameter. Most varieties available are seed-free.

MAJOR PRODUCERS:

The major producers of lime are Mexico, U.S.A., India and Jamaica.

STANDARDS:

There are no international standards for limes. There are many recommendations, e.g. U.S. Grade Standards, but they are not mandatory in international trade.

MINIMUM REQUIREMENTS:

Lime should be intact, clean, sound and free from attack by disease, insects, rot or mould. They should be firm and the rind without spots or damage. The flesh should be juicy and without chilling injury, which may result in dry, discoloured flesh.

Latin: Citrus aurantiifolia (Christm.) Swingle
French: Lime or Limette
German: Limette or Limone
Spanish: Lima or Limón dulce

LIME

KEEPING QUALITIES:

Shelf life:

10°C, 90% RH, 6-8 weeks
20°C, 60% RH, 2 weeks

Ideal demands: 8-10°C (46-50°F), 90% RH

Recommended temperature:

0	4	8	12	16	°C
32	39	46	54	61	°F

Sensitivity:

T	RH	ETHYLENE PROD.	SENS.	CHILLING INJURIES
★	★	★	★	★★

Heat production:

HEAT PRODUCTION
W/ton

Specific heat: 3.72 kJ/kg x °C 0.89 Btu/lb x °F
Specific weight: Boxes, pallets 350-450 kg/m^3. Bulk 480-520 kg/m^3

157

LOQUAT (Japanese medlar)

Latin: Eriobotrya japonica (Thunb.) Lindl.
French: Néfle du Japon
German: Japanische Mispel or Loquat
Spanish: Nispola del Japón or Nispero

DESCRIPTION OF PRODUCT:

Loquat has its origin in South East Asia, in all probability south-eastern China. From there it spread to Japan where it was hybridized and cultivated several thousand years before it again spread to many other parts of the world. Now it is an ordinarily cultivated plant in all parts of the world.

Loquat is a small evergreen tree with lance-shaped leaves with a dark green surface and a downy, yellowish grey underside. The flowers, which grow upright in a tuft, are small, white and have a strong fragrance.

The fruit is round or oval and is the size of a plum. The skin is yellow or orange, depending on the variety, and can be rather tough. The juicy flesh is yellowish or orange and contains a parchment-like core containing up to 9 seeds. Loquat has a mild and sweet taste, perhaps a little sourish, but not particularly exotic.

MAJOR PRODUCERS:

The major producers of loquat are Italy, Japan, Israel, Brazil and U.S.A.

STANDARDS:

There are no international standards for loquat.

MINIMUM REQUIREMENTS:

Loquat should be intact, clean, sound and fresh. They should be full-bodied and the skin should not be tough. There should be no spots or indication of mechanical damage. Depending on the variety loquat should be yellow or orange, and the flesh pale without any dark parts and sign of attack by disease, rot or mould.

Latin: Eriobotrya japonica (Thunb.) Lindl.
French: Nêfle du Japon
German: Japanische Mispel or Loquat
Spanish: Nispola del Japón or Nispero

(Japanese medlar) **LOQUAT**

KEEPING QUALITIES:

Shelf life:

0°C, 90% RH, 2-3 weeks
20°C, 60% RH, 3-5 days

Ideal demands: 0°C (32°F), 90–95% RH

Recommended temperature:

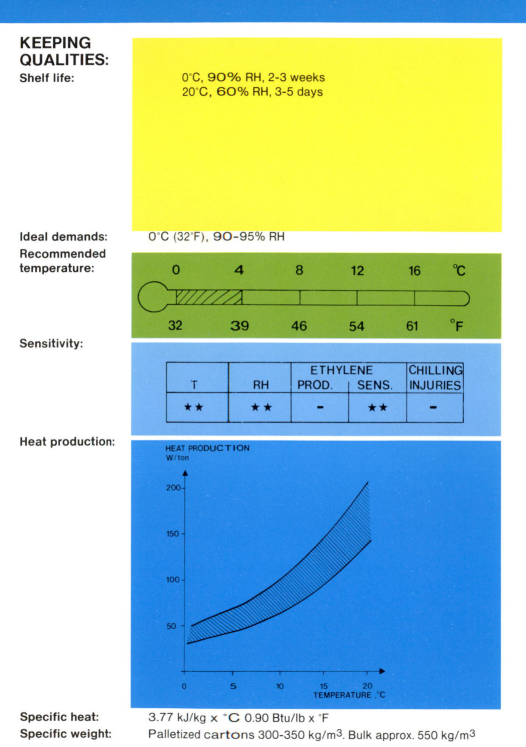

| 0 | 4 | 8 | 12 | 16 | °C |
| 32 | 39 | 46 | 54 | 61 | °F |

Sensitivity:

T	RH	ETHYLENE PROD.	SENS.	CHILLING INJURIES
★★	★★	–	★★	–

Heat production:

HEAT PRODUCTION
W/ton

TEMPERATURE ,°C

Specific heat: 3.77 kJ/kg x °C 0.90 Btu/lb x °F

Specific weight: Palletized cartons 300-350 kg/m³. Bulk approx. 550 kg/m³

DESCRIPTION OF PRODUCT:

Lychee originates from China, where it has been cultivated for at least 1000 years. Cultivation spread from eastern Asia, but as the lychee tree is difficult to grow it took a long time before cultivation outside China and India became wide-spread.

The lychee is an evergreen tree of average height. The fruit is egg-shaped, approx. 3 cm in diameter, and has a strong, leathery, red skin which is bluntly prickled. Underneath is the edible white flesh which encloses a large seed. The fruit is prone to rot and mould.

Rambutan (Latin: Nephelium lappaceum) is native to Malaysia and belongs to the same family as lychee. Its skin is co-vered with hair and the flesh is slightly yellow.

MAJOR PRODUCERS:

The major producers of lychee are China, India, South Africa, Australia and U.S.A.

STANDARDS:

There are no international standards for lychee.

MINIMUM REQUIREMENTS:

Lychee should be intact, clean, have a fresh appearance without visible signs of attack by disease, pests, rot or mould. The fruit should be free of cracks and damage of any sort. The flesh should be firm or slightly jellyish and white. Lychee should not emit any foreign taste or smell.

Latin: Litchi chinensis Sonn.
French: Litchi
German: Litchi
Spanish: Litchi or Litco

LYCHEE

KEEPING QUALITIES:

Shelf life:

Lychee:	5°C, 90% RH, 4-6 weeks
	20°C, 60% RH, 7-10 days
Rambutan:	10°C, 90% RH, 1-2 weeks
	20°C, 60% RH, 3-5 days

Ideal demands:

Lychee: 2-6°C (36-43°F), 90-95% RH Rambutan: 10°C (50°F), 90-95% RH

Recommended temperature:

0	4	8	12	16	°C
32	39	46	54	61	°F

Sensitivity:

T	RH	ETHYLENE PROD.	SENS.	CHILLING INJURIES
★★	★	★	-	★★

Heat production:

This diagram is for lychee. Heat production of rambutan is 1 1/2 to 2 times higher.

HEAT PRODUCTION
W/ton

200 —

150 —

100 —

50 —

0   5   10   15   20
TEMPERATURE ,°C

Specific heat: 3.60 kJ/kg x °C 0.86 Btu/lb x °F

Specific weight: Lychee: Boxes, pallets 250-280 kg/m³.
Rambutan: Boxes, pallets 220-260 kg/m³

DESCRIPTION OF PRODUCT:

Mango is said to have come from the mountain regions of Burma and the Himalayas. Mango is known to have been cultivated in India several thousands of years before the Christian era. From there it spread to South East Asia, Malaysia and China among others, where descriptions show that cultivation of mangoes was then widespread. Mangoes, next after bananas, is the second most important tropical fruit cultivated in the world.

The mango tree is tall, about 40 m, and has a vast crown. It has top-shaped inflorescence each with several thousand flowers.

Mangoes may be round, oval, pear or kidney-shaped and the colour of the skin green, or ranging from yellow-green to red, orange and violet. The flesh is yellow-orange and in some varieties slightly thready. It contains a large, flat and rather thready seed which is firmly embedded in the flesh. The fruit tastes sweet and is very aromatic.

Some mangoes can have a predominant turpentine-like taste.

MAJOR PRODUCERS:

India is by far the largest producer of mangoes, but Mexico, Pakistan, Indonesia and Brazil are also important producers.

STANDARDS:

There are no international standards for mangoes.

MINIMUM REQUIREMENTS:

Mangoes should be clean, intact, sound and fresh in appearance without signs of attack by disease or pests. They should be free from spots, mechanical damage, pressure marks and bruises and from foreign smell or taste. Ripe mangoes yield to slight pressure. Most ripe mangoes are yellow or red, though some varieties may remain green. The flesh of a ripe fruit is yellow. Mangoes damaged by chilling have fibrous flesh and smell of turpentine.

Latin: Mangifera indica L.
French: Mangue
German: Mango (frucht)
Spanish: Mango

KEEPING QUALITIES:

Shelf life:

12°C, 85% RH, 2-3 weeks
20°C, 60% RH, 3-4 days

Ideal demands: 10-14°C (50-57°F), 85-90% RH

Recommended temperature:

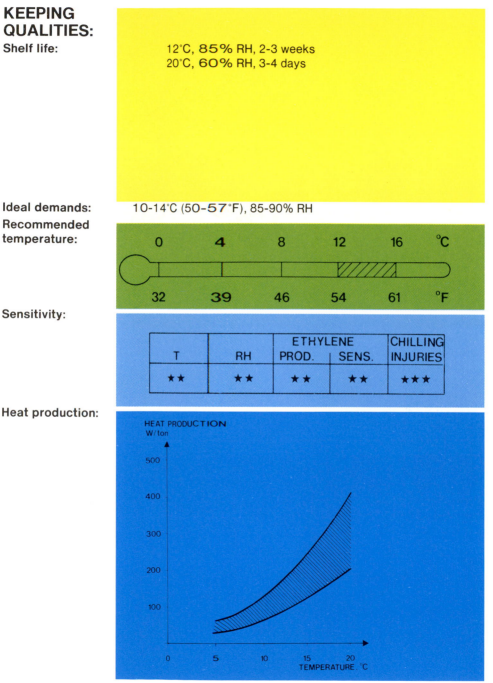

0	4	8	12	16	°C
32	39	46	54	61	°F

Sensitivity:

T	RH	ETHYLENE PROD.	SENS.	CHILLING INJURIES
★★	★★	★★	★★	★★★

Heat production:

HEAT PRODUCTION
W/ton

500

400

300

200

100

0   5   10   15   20

TEMPERATURE . °C

Specific heat: 3.56 kJ/kg x °C 0.85 Btu/lb x °F
Specific weight: Boxes, pallets 280-350 kg/m³

MANGOSTEEN

Latin:	Garcinia mangostana L.
French:	Mangoustan
German:	Mangostane
Spanish:	Mangostán

DESCRIPTION OF PRODUCT:

Mangosteen is native to Malaysia where it has been cultivated for many years. It grows in the tropical rain forest and is very difficult to cultivate. It is therefore not widespread outside South East Asia, although it is considered to be one of the finest tropical fruits.

The mangosteen tree, which grows up to 15 m, is an evergreen with short stalked, oblong leaves. The fruit is 6-8 cm in diameter. The shell, approx. 0.5 cm thick, is red or reddish brown, hard and leathery. It has 5 thick sepals.

There are 4–7 seeds enclosed in white or pink flesh. Each segment resembles a clove of garlic in shape. The flesh is soft and juicy, the taste sweet and slightly spicy. In spite of its hard shell it can be opened by hand by pressing the fruit's "equator".

MAJOR PRODUCERS:

The major producers of mangosteen are Thailand, Malaysia, Indonesia and Brazil.

STANDARDS:

There are no international standards for mangosteen.

MINIMUM REQUIREMENTS:

Mangosteen should be intact and sound. They should bear no growth cracks or other deep injuries that may damage the flesh. The colour should be brown-violet. The fruit should be dry and firm without any soft spots. The flesh should be white and without discolouration, juicy and sweet. It should not taste or smell unpleasant nor bear signs of attack by disease, rot or pests.

KEEPING QUALITIES:

Shelf life:

5°C, 90% RH, 6-7 weeks
20°C, 60% RH, 2-3 weeks

Ideal demands: 4-6°C (39-43°F), 85-90% RH

Recommended temperature:

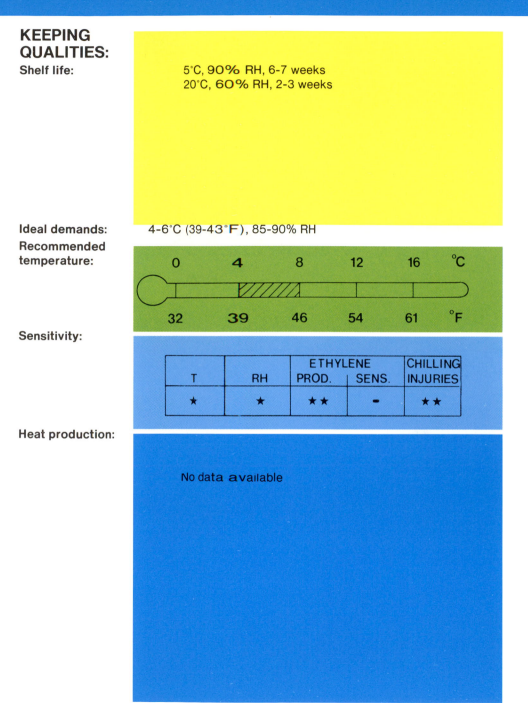

	0	4	8	12	16	°C
	32	39	46	54	61	°F

Sensitivity:

T	RH	ETHYLENE PROD.	SENS.	CHILLING INJURIES
★	★	★★	-	★★

Heat production:

No data available

Specific heat: 3.56 kJ/kg x °C 0.85 Btu/lb x °F

Specific weight: Boxes, pallets 280-340 kg/m³

MANIOC

Latin:	Manihot esculenta Cranz.
French:	Manioc
German:	Manioc, Kasava or Tapioca
Spanish:	Guacomote or yuca

DESCRIPTION OF PRODUCT:

Manioc is native to South America where it has been an important cultivated plant for almost 5000 years. From here it was brought to Africa and not until much later did it spread to eastern Asia. Now it is cultivated in large parts of the tropical region.

The manioc plant is an herbaceous perennial which is bush-like and grows to a height of up to 5 m. The leaves are dissected. The tubers are swollen roots containing food reserves. They can weigh up to 10 kg, but normally they are about 2 kg. The skin is a thin layer of cork under which there is a bark layer containing starch and, thereafter, a woody layer also containing starch. The phloem, which lies between the bark and the xylem contains a glucoside, linamarin. This glucoside converts easily to prussic acid which is very poisonous. Careful boiling or frying can remove the glucoside and prussic acid and therefore, manioc cannot be eaten raw.

MAJOR PRODUCERS:

The major producers of manioc are Brazil, Zaire, Thailand, Nigeria and Indonesia.

STANDARDS:

There are no international standards for manioc.

MINIMUM REQUIREMENTS:

Manioc should be clean and sound. There should be no signs of attack by disease, pests, rot or mould. The surface should be smooth and undamaged.

Latin:	Manihot esculenta Cranz.
French:	Manioc
German:	Manioc, Kasava or Tapioca
Spanish:	Guacomote or yuca

MANIOC

KEEPING QUALITIES:

Shelf life:

1°C, 90% RH, 5-6 months
20°C, 60% RH, 2-4 weeks

Ideal demands: 0-2°C (32-36°F), 85-90% RH

Recommended temperature:

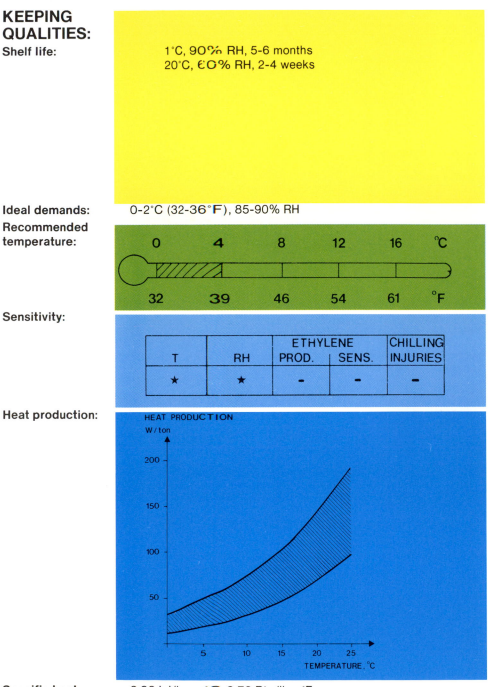

	T	RH	ETHYLENE PROD.	SENS.	CHILLING INJURIES
	★	★	-	-	-

Sensitivity:

Heat production:

HEAT PRODUCTION
W/ton

Specific heat: 2.93 kJ/kg x °C 0.70 Btu/lb x °F

Specific weight: Boxes, pallets 250-350 kg/m³. Bulk approx. 450 kg/m³

167

MELON

DESCRIPTION OF PRODUCT:

The melon probably is native to eastern India where it has been cultivated for thousands of years.

It is an annual vine whose very watery fruits contain a profusion of seeds. There is a vast variety of melons and just as many shapes, one of the reasons being that they are cultivated all over the world and, therefore, under widely differing conditions from country to country. The following types of melon are among the most common:

Net Melon is rather large and round with a characteristic scab-like meshed surface. It is yellow.

Honeydew Melon is smaller than a netted melon and has longitudinal grooves in the skin. Some early varieties are yellow, while certain late sorts have a dark rind.

Cantaloupe has diameter of 10-15 cm and is smaller than the honeydew which it closely resembles.

Ogen Melon is the smallest type. It is round with stripes from stalk to blossom end. It is yellow-green.

MAJOR PRODUCERS:

The major producers of melons are China, U.S.A., Spain, Iran and Egypt.

STANDARDS:

There are no international standards for melons. However, there are many recommendations, but they are not mandatory in international trade.

MINIMUM REQUIREMENTS:

Melons should be whole, fresh and full-bodied. They should be clean and uninfected by disease, rot or insects. Melons should be well-shaped, typical of the variety, and of a suitable degree of ripeness. They should be free from signs of mechanical damage, spots or discolouration. The flesh should be juicy, firm without being fibrous or woody, aromatic and of a colour typical of the variety.

KEEPING QUALITIES:
Shelf life:

Net Melons:	6-9°C, 85-90% RH, 10-14 days
	20°C, 60% RH, 3-5 days
Honey Melons:	10-14°C, 85-90% RH, 16-20 days
	20°C, 60% RH, 10-14 days
Cantaloupes:	3-5°C, 85-90% RH, 10-14 days
	20°C, 60% RH, 3-5 days
Ogen Melons:	6-7°C, 85-95% RH, 10-14 days
	20°C, 60% RH, 2-3 days
Watermelons:	5-6°C, 80-85% RH, 16-20 days
	20°C, 60% RH, 8-12 days

Ideal demands: 5-9°C (41-48°F), 85-95% RH, depending on the variety.

Recommended temperature:

0	4	8	12	16	°C
32	39	46	54	61	°F

Sensitivity:

T	RH	ETHYLENE PROD.	SENS.	CHILLING INJURIES
★★	★★	★★	★	★★

Heat production:

HEAT PRODUCTION
W/ton

150

100

50

0   5   10   15   20
TEMPERATURE . °C

Specific heat: 4.01 kJ/kg x °C 0.96 Btu/lb x °F

Specific weight: Boxes, pallets 220-430 kg/m^3. Bulk approx. 600 kg/m^3

MUSHROOM

DESCRIPTION OF PRODUCT:

A mushroom is a fungus, a so-called spore plant, and is, therefore, not a vegetable although it is ordinarily referred to as one. Mushrooms, which claim a large number of improved varieties, differ from vegetables as they are unable to produce their own organic matter, i.e. they feed on left-over organic matter on host plants.

Mushrooms grow in woods and fields in large parts of the world, but it was around the 17th century in France that a method was found to develop a mushroom culture. Mushroom cultivation has now spread all over the world and mushrooms are the most common edible fungus.

The edible part of the mushroom is the fruiting body which is the visible part where spores are produced. The large, white, closed cap with unexposed gills is often preferred. However, there are many other varieties such as brown mushrooms which are popular in many places.

MAJOR PRODUCERS:

The major producers of mushrooms are U.S.A., France, China, the Netherlands and England.

STANDARDS:

There are no international standards for mushrooms, but there are many recommendations, e.g. EEC Standard No. FFV-24 and U.S. Grade Standards, but they are not mandatory in international trade.

MINIMUM REQUIREMENTS:

Mushrooms should be pale, almost white, without brown spots. They should be closed, intact, full-bodied, and have firm caps and stems and almost no adhering soil. The roots should show no signs of mycelium growth. The stems should show no discolouration and the veil should be intact. If the roots have been removed the cut edges should be free of soil. They should be free from disease and internal brown discolouration.

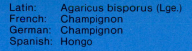

Latin:	Agaricus bisporus (Lge.)
French:	Champignon
German:	Champignon
Spanish:	Hongo

KEEPING QUALITIES:
Shelf life:

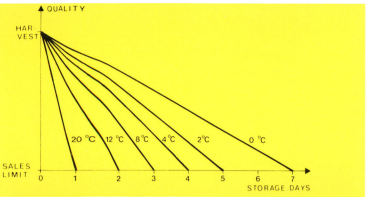

Ideal demands: 0°C (32°F), 90–95% RH

Recommended temperature:

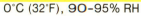

Sensitivity:

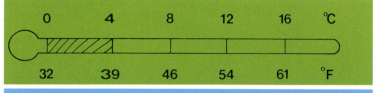

	T	RH	ETHYLENE PROD.	SENS.	CHILLING INJURIES
	★★★	★★★	★★	−	−

Heat production:

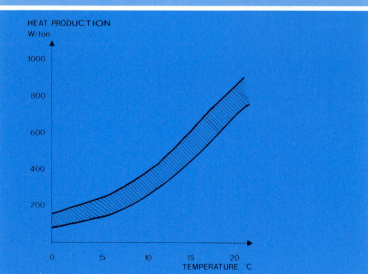

Specific heat: 3.91 kJ/kg x °C 0.93 Btu/lb x °F

Specific weight: Palletized cartons 160-230 kg/m³. Bulk 250-500 kg/m³ (dep. on size)

171

OKRA (Ladies finger or gombo)

Latin:	Hibiscus esculentus L.
French:	Gombo
German:	Gombo or Okra
Spanish:	Quingombo

DESCRIPTION OF PRODUCT:

Okra is believed to have come from tropical Africa, but was also cultivated in certain parts of Asia for a long time.

The plant which can grow to a height of 2 m is an annual herb with big lobed leaves. The flowers are self-fertile.

The fruit, which resembles a finger, is a capsule with lengthwise furrows. There are many small, slippery seeds on the long placenta inside. The capsule is green and often downy. Chilling injury results in superficial discolouration and spots.

Cooking releases a slimy juice which often works as a thickening agent.

MAJOR PRODUCERS:

Okra is produced in tropical regions, especially India, Thailand, Kenya and Central America.

STANDARDS:

There are no international standards for okra. However, there are many recommendations, e.g. U.S. Grade Standards, but they are not mandatory in international trade.

MINIMUM REQUIREMENTS:

Okra should be intact, clean and sound. They should be free from attack by disease, insects, rot and mould. Okra should be full-bodied and the colour a fresh green. They should be free from damage, discolouration and spots. The seeds should not be coarse.

172

KEEPING QUALITIES:

Shelf life:

10°C, 90% RH, 7-10 days
20°C, 60% RH, 1-2 days

Ideal demands: 8-10°C (46-50°F), 90-95% RH

Recommended temperature:

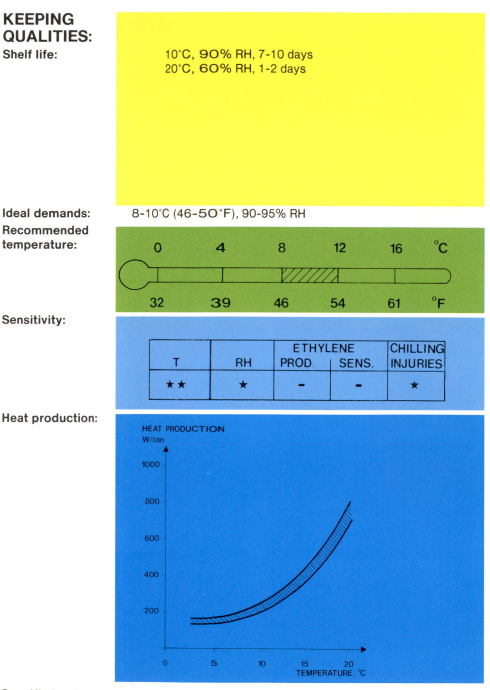

Sensitivity:

T	RH	ETHYLENE PROD.	SENS.	CHILLING INJURIES
★★	★	−	−	★

Heat production:

Specific heat: 3.85 KJ/kg × °C 0.92 Btu/lb × °F

Specific weight: Boxes, pallets 240-280 kg/m³. Bulk approx. 400 kg/m³

173

DESCRIPTION OF PRODUCT:

Onions originate from Central Asia. Wild Kepa onions are still to be found in North-West China, Afghanistan and Uzbekistan. Onions are composed of juicy, tight, concentric layers. When harvested the stem and the outer skin are dried, this closes the onion and protects it against diseases.

MAJOR PRODUCERS:

The major producers of onions are China, India, U.S.S.R., U.S.A. and Japan.

STANDARDS:

For trade within the EEC onions must fulfil EEC Standard No. 4. There are many recommendations, e.g. U.S. Grade Standards, but they are not mandatory in international trade.

MINIMUM REQUIREMENTS:

Onions should be whole, sound, clean, free of damage due to frost and free of extraneous foreign matter. Onions should be firm and well dried with no foreign taste or smell. The outer scales should not be missing. The surface should be moist and the stem should not exceed 4 cm. There should be no signs of sprouting or re-growth.

Latin: Allium cepa (L.)
French: Oignon
German: Zwiebel
Spanish: Cebolla

KEEPING QUALITIES:
Shelf life:

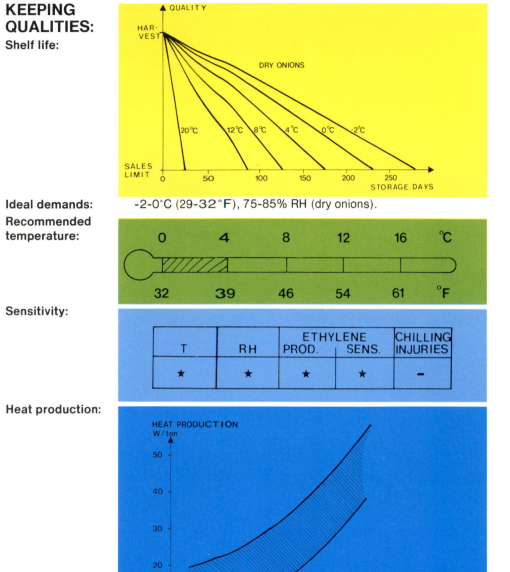

Ideal demands: -2-0°C (29-32°F), 75-85% RH (dry onions).

Recommended temperature:

0	4	8	12	16	°C
32	39	46	54	61	°F

Sensitivity:

T	RH	ETHYLENE PROD.	SENS.	CHILLING INJURIES
★	★	★	★	–

Heat production:

Specific heat: 3.78 kJ/kg x oC 0.90 Btu/lb x °F
Specific weight: Palletized sacks 350-420 kg/m³. Bulk approx. 550 kg/m³

DESCRIPTION OF PRODUCT:

The orange is assumed to have its origin in southern China. It was brought westward to India and to southern Europe where it has been cultivated since the fifteenth century. Orange trees were then mostly ornamental.The first large orange plantation was cultivated no earlier than the end of the eighteenth century.

The orange tree is between 5 and 12 metres tall and is usually very thorny. The fruit is generally round and has a tough, leathery rind that protects the flesh from drying up.

Each part of the fruit is utilized. The flesh is pressed for its juice, the rind con tains orange oil used in soft drinks, liqueurs and cosmetics, the pith is used in the production of pectin and the dried remains of the rind after oil extraction are used as feedstuff.

MAJOR PRODUCERS:

The major producers of oranges are Brazil, U.S.A., Italy and China.

STANDARDS:

For trade within the EEC oranges must comply with EEC Standard No. 18. There are many additional recommendations, but they are not mandatory in international trade.

MINIMUM REQUIREMENTS:

Oranges should be whole, sound, free from damage by frost and free from visible foreign matter. Oranges in a package should be of the same variety. There should be no trace of foreign smell or taste. Oranges should contain min. 35% juice of the fruit's total weight.

Latin: Citrus sinensis (L.) Osbeck
French: Orange douce
German: Apfelsine
Spanish: Naranja

ORANGE

KEEPING QUALITIES:

Shelf life:

Storing time and temperature depends on the variety and the country of origin.

Examples:

Navel,	Spain:-3°C, 8-10 weeks
Navel,	California:2-7°C, 5-8 weeks
Valencia,	Florida: 0-1°C, 8-12 weeks
Shamouti,	Israel:4-5°C, 6-8 weeks

Ideal demands:

0-5°C (32-41°F), 85-90% RH, dep. on the var. and the country of origin.

Recommended temperature:

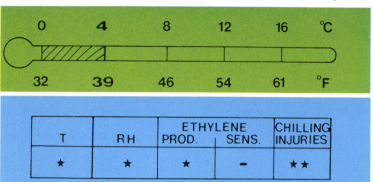

Sensitivity:

T	RH	ETHYLENE PROD.	SENS.	CHILLING INJURIES
★	★	★	–	★★

Heat production:

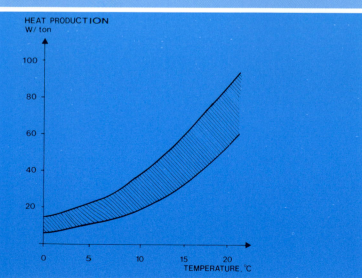

Specific heat: 3.77 kJ/kg x °C 0.90 Btu/lb x °F

Specific weight: Palletized cartons 350-450 kg/m³. Bulk 480-520 kg/m³

PAK CHOI (Celery cabbage)

Latin:	Brassica chinensis L.
French:	Pe-tsai or pak-choi
German:	Paksoi or pak-choi
Spanish:	Pak choy

DESCRIPTION OF PRODUCT:

Pak choi is of Asiatic origin. In China and Japan this vegetable has been cultivated for many years, but only the last 10 years have shown an increasing interest from the rest of the world.

Pak choi resembles a leafy beet. The leaves are green-dark green, the mid-ribs and the leaf stalks are thick, fleshy and white – a decorative plant. The taste is very similar to that of Chinese cabbage to which it is closely related.

Pak choi is adapted to grow in a salty environment and demands shelter and good growth conditions. However, new varieties which are adapted to various climatic conditions are on the way.

MAJOR PRODUCERS:

The major producers of pak choi are China, Japan, Korea and Taiwan.

STANDARDS:

There are no international standards for pak choi.

MINIMUM REQUIREMENTS:

Pak choi should be fresh, intact and clean. Leaf stalks should sit closely together to form a whole. Leaves and stalks should be green and full-bodied without signs of withering. The stalks should be pale and tender and bear no spots or signs of mechanical damage or of attack by disease, pests, rot or mould.

Latin: Brassica chinensis L.
French: Pe-tsai or pak-choi
German: Paksoi or pak-choi
Spanish: Pak choy

(Celery cabbage) **PAK CHOI**

KEEPING QUALITIES:

Shelf life:

0°C, **90%** RH, 30-40 days
20°C, **60%** RH, 2-3 days

Ideal demands: 0°C (32°F), 90-95% RH

Recommended temperature:

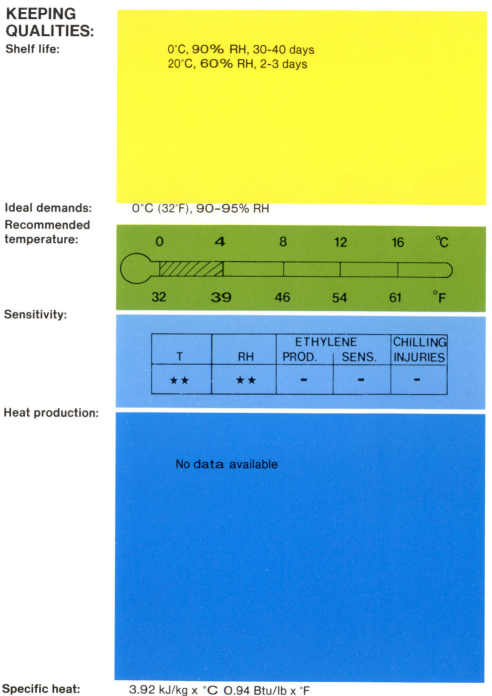

| 0 | 4 | 8 | 12 | 16 | °C |
| 32 | 39 | 46 | 54 | 61 | °F |

Sensitivity:

T	RH	ETHYLENE PROD.	SENS.	CHILLING INJURIES
★★	★★	-	-	-

Heat production:

No data available

Specific heat: 3.92 kJ/kg x °C 0.94 Btu/lb x °F

Specific weight: Boxes, pallets **1** 50-200 kg/m³. Bulk approx. 220 kg/m³

179

DESCRIPTION OF PRODUCT:

Papaya is believed to have come originally from Mexico. The plant is a palm-like tree. The fruits hang closely together around the stem. The fruit is pear-shaped, and can be up to 30 cm long. It has a thin rind, juicy flesh and is hollow in the centre where its many inedible seeds lie.

Size and colour of the skin and flesh vary from variety to variety. Ripe papaya are either yellow or green. The flesh in most cases is orange or yellowish.

The fruits should be full-bodied without any foreign smell or taste. Papayas should be well developed and without deformities. Ripe papayas are yellow or yellow-green and yields to a slight pressure. The even coloured flesh is as soft as that of melons. Papayas with chilling injury have large, hollow spots on the rind and they seldom ripen.

MAJOR PRODUCERS:

The major producers of papaya are Brazil, India, Indonesia, Zaire and Colombia.

STANDARDS:

There are no international standards for papayas.

MINIMUM REQUIREMENTS:

Papayas should be intact, clean, sound, fresh in appearance, show no signs of attack by disease or pests and be free from spots, growth cracks, mechanical damage or bruises from rough handling.

Latin: Carica papaya L.
French: Papaye
German: Papayafrucht or Baummelone
Spanish: Papaya

(Paw paw) **PAPAYA**

KEEPING QUALITIES:

Shelf life:

10°C, 90% RH, 2-3 weeks
20°C, 60% RH, 2-3 days

Ideal demands: 10°C (50°F), 85–90% RH

Recommended temperature:

| 0 | 4 | 8 | 12 | 16 | °C |
| 32 | 39 | 46 | 54 | 61 | °F |

Sensitivity:

T	RH	ETHYLENE PROD.	SENS.	CHILLING INJURIES
★★	★★	★★★	★★	★★★

Heat production:

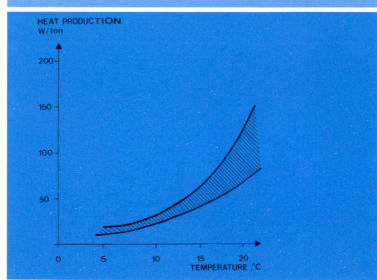

HEAT PRODUCTION
W/ton

Specific heat: 3.89 kJ/kg x °C 0.93 Btu/lb x °F

Specific weight: Boxes, pallets 260-320 kg/m³

PARSNIP

Latin: Pastinaca sativa L.
French: Panais
German: Pastinake
Spanish: Pastinaca

DESCRIPTION OF PRODUCT:

Parsnip is native to southern and central
Europe where they grow especially in
marshy regions. The original form is still
found growing wild.

The parsnip is a very old cultivated plant
which has been known for several thousands of years.

There are many varieties of parsnip:
long, conical or round. The long variety is
most common and it can grow to 40 cm
in length.

Parsnips can withstand some frost and
are, therefore, capable of wintering outdoors in many places.

MAJOR PRODUCERS:

The major producers of parsnips are
U.K., West Germany and the Netherlands.

STANDARDS:

There are no international standards for
parsnips.

MINIMUM REQUIREMENTS:

Parsnips should be intact, sound, clean,
full-bodied and fresh-looking. They
should be well shaped without branching, free from attack by disease and
pests, rust or other discolourations with
no signs of mechanical damage or
cracks and should not be woody. If parsnips are sold with leaves, they should
be fresh and green.

KEEPING QUALITIES:

Shelf life:

0°C, 95% RH, 2-6 months
20°C, 60% RH, 4-6 days

Ideal demands: 0°C (32°F), 90–95% RH

Recommended temperature:

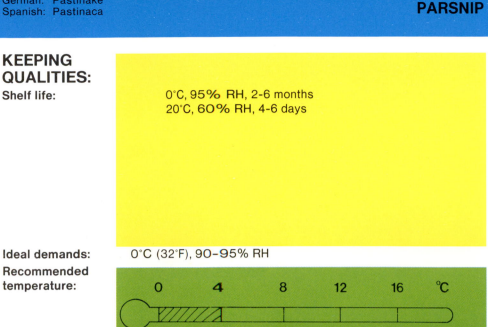

Sensitivity:

T	RH	ETHYLENE PROD.	SENS.	CHILLING INJURIES
★	★★	–	–	–

Heat production:

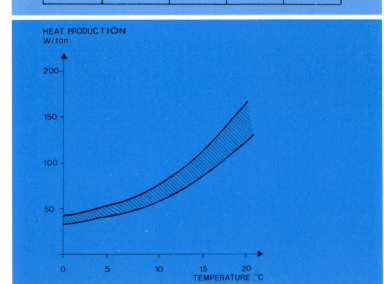

Specific heat: 3.47 kJ/kg x °C 0.83 Btu/lb x °F

Specific weight: Boxes, pallets 280-350 kg/m³. Bulk approx. 250 kg/m³

DESCRIPTION OF PRODUCT:

Passion fruit is native to South America where it is widespread in the subtropical regions and tropical highlands. From there it spread to the rest of the world and is now cultivated in all continents.

This vine is a lignified perennial which may grow to a length of 15 m. It has spiral tendrils. The flowers are white-violet and, due to their attractive appearance, the plant is often used as an ornament.

Passion fruits are thick-shelled berries which contain a number of seeds. Each seed is enclosed in an aril. These arils collectively comprise the jelly-like edible flesh. Passion fruits have a refreshing, somewhat sour taste and are aromatic.

There are many varieties. The violet or purple passion fruit is a small round fruit with a wrinkled skin. The flesh is yellow-green. The yellow passion fruit (sweet or yellow grenadilla) is bigger and more or less oval. There are several varieties of this type on the market including a light yellow variety with a slightly wrinkled shell and yellowish flesh and a darker yellow red-cheeked variety with a smooth hard shell and greenish flesh.

MAJOR PRODUCERS:

The major producers of passion fruits are Brazil, South Africa, U.S.A., Australia and Kenya.

STANDARDS:

There are no international standards for passion fruits.

MINIMUM REQUIREMENTS:

Passion fruits should be intact, clean, sound and without signs of attack by disease or insects. The fruit should be free from mechanical damage. The skin should show no signs of cracks or spots. The fruits wrinkle easily, but they should not dry out.

184

Latin:	Passiflora edulis Sims
French:	Grenadille
German:	Grenadil or Passionsfrucht
Spanish:	Granadilla

PASSION FRUIT (Grenadilla)

KEEPING QUALITIES:

Shelf life:

8°C, 90% RH, 3-4 weeks
20°C, 60% RH, approx. 1 week

Ideal demands: 7-10°C (45-50°F), 85-90% RH

Recommended temperature:

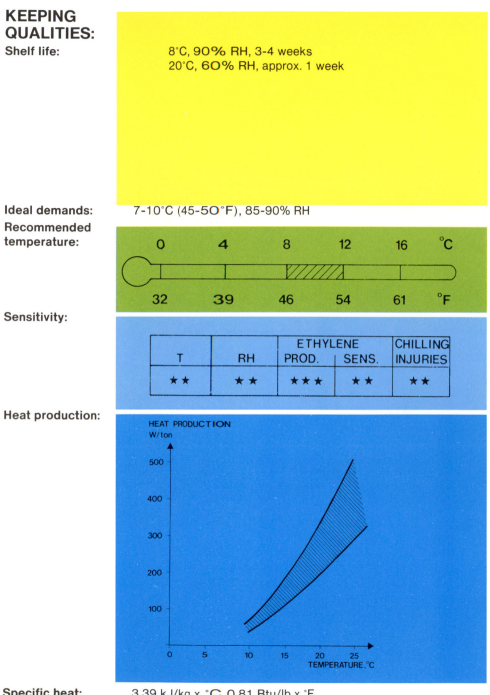

Sensitivity:

T	RH	ETHYLENE PROD.	SENS.	CHILLING INJURIES
★★	★★	★★★	★★	★★

Heat production:

Specific heat: 3.39 kJ/kg x °C 0.81 Btu/lb x °F
Specific weight: Boxes, pallets 220-280 kg/m³

185

PEA

DESCRIPTION OF PRODUCT:

Peas originate from the eastern Mediterranean region, Iran, Afghanistan and Tibet. They have been known and cultivated for thousands of years.

The pea, an annual vine, 30-150 cm long, has a stem with pinnate leaves and terminal branching tendrils. The most commonly cultivated variety for fresh consumption is the marrowfat which is distinctively large and soft. The pod has a tough membrane.

The size of the pods depends on the variety. An average pod has a 7-10 cm length with 5-10 green peas which may be 8-13 mm in diameter.

Peas for fresh consumption have to be harvested at the correct time. If picked too early the pods will not be full and if harvested too late the green peas will be hard and mealy.

MAJOR PRODUCERS:

The major producers of green peas are U.S.A., U.K., France, Hungary and China.

STANDARDS:

For trade within the EEC green peas must comply with EEC Standard No. 15. There are, however, many recommendations, e.g. U.S. Grade Standards, but they are not mandatory in international trade.

MINIMUM REQUIREMENTS:

Green peas in pods should be juicy, firm and crisp. They should not be mealy. The pods should be intact, clean and sound, free from any signs of attack by disease or pests, mould or rot. The pods should be well developed, full and free from damage.

KEEPING QUALITIES:
Shelf life:

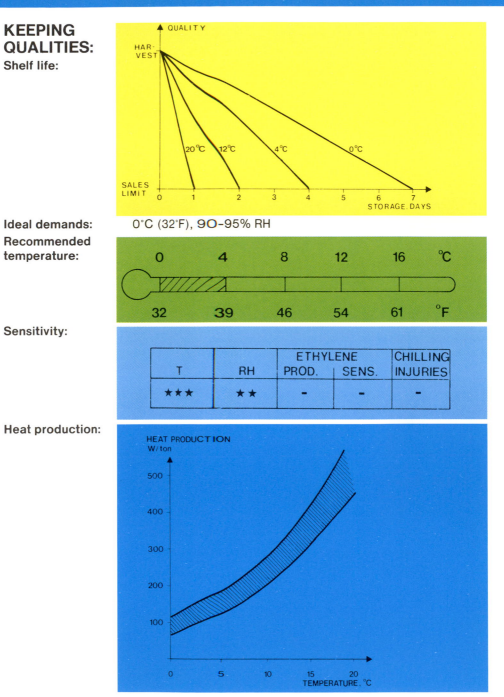

		ETHYLENE		CHILLING
T	RH	PROD.	SENS.	INJURIES
★★★	★★	-	-	-

Ideal demands: 0°C (32°F), 90-95% RH

Recommended temperature:

Sensitivity:

Heat production:

Specific heat: 3.74 kJ/kg x °C 0.89 Btu/lb x °F

Specific weight: Boxes, pallets 150-190 kg/m³. Bulk approx. 250 kg/m³

PEACH AND NECTARINE

Latin: Prunus persica (L.) Batsch
French: Pêche and Nectarine
German: Pfirsich and Nektarine
Spanish: Durazno or persico

DESCRIPTION OF PRODUCT:

It is believed that peaches originate from China where they have been cultivated for at least 4000 years. They spread from there to Japan and via caravan routes over the Himalayas to Persia.

The peach tree can be 5 m tall and has slender leaves. The fruit is a stone fruit with yellow, juicy flesh. The stone has a perforated surface and contains a kernel which is often used in the production of marzipan substitutes and other similar products.

There are a few thousand varieties of peach. Most of them are downy, but there are some smooth-skinned varieties which are called nectarines.

MAJOR PRODUCERS:

The major producers of peaches and nectarines are Italy, U.S.A., Greece, Spain and U.S.S.R.

STANDARDS:

For trade within the EEC peaches must comply with EEC Standard No. 6b. There are many other recommendations, e.g. U.S. Grade Standards, but they are not mandatory in international trade.

MINIMUM REQUIREMENTS:

Peaches should be intact, sound, clean, and free from rot or mould and any foreign smell or taste. There should be no signs of mechanical damage. The fruits must be full-bodied without any bruises or brown discolouring of the flesh.

Latin: Prunus persica (L.) Batsch
French: Péche and Nectarine
German: Pfirsich and Nektarine
Spanish: Durazno or persico

PEACH AND **NECTARINE**

KEEPING QUALITIES:
Shelf life:

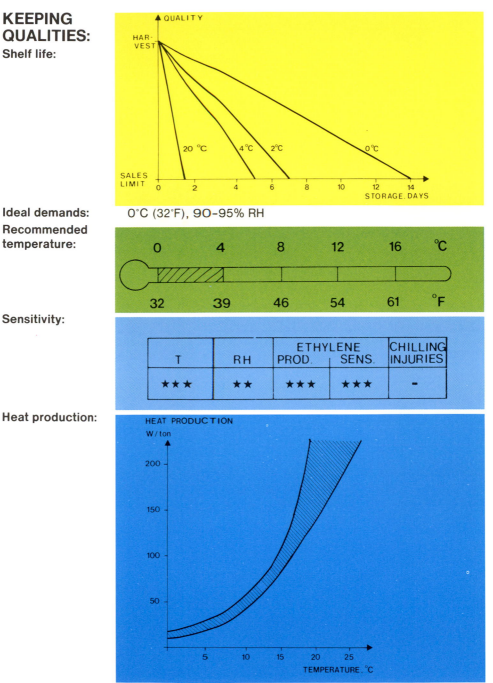

Ideal demands: 0°C (32°F), 90–95% RH

Recommended temperature:

0	4	8	12	16	°C
32	39	46	54	61	°F

Sensitivity:

T	RH	ETHYLENE PROD.	SENS.	CHILLING INJURIES
★★★	★★	★★★	★★★	–

Heat production:

HEAT PRODUCTION
W/ton

Specific heat: 3.81 kJ/kg x °C 0.91 Btu/lb x °F
Specific weight: Palletized boxes 210-280 kg/m³. Bulk approx. 610 kg/m³

PEAR

Latin: Pyrus Communis L.
French: Poire
German: Birne
Spanish: Pera

DESCRIPTION OF PRODUCT:

It is uncertain where pears originate from. Wild pears can be found in several parts of Asia and Europe, but a very thorny pear tree from north-west Asia is believed to be the most probable original form.

There are numerous varieties with many shapes and sizes, colours, tastes and aromas. Pears are usually 45-80 mm in diameter. The flesh is pale, very juicy and easily bruised. Pressure marks result in brown discolouration.

MAJOR PRODUCERS:

The major producers of pears are China, Italy, U.S.A., U.S.S.R. and West Germany.

STANDARDS:

For trade within the EEC pears must comply with EEC Standard No. 1. There are many recommendations, e.g. U.S. Grade Standards, but they are not mandatory in international trade.

MINIMUM REQUIREMENTS:

Pears should be intact, clean, sound and full-bodied. They should be free from any visible foreign matter and from foreign taste or smell. Skin and flesh should be free from spots. There should be no signs of attack by disease, pests, mould or rot.

KEEPING QUALITIES:

Shelf life:

Pears are kept at -0.5-0°C and may be stored 1-6 months, depending on the variety and storage method.

Examples:

Cold storage: Bartlett - 2.5-3 months
Anjou - 4-6 months

C.A.: Conference - 5-6 months
Doyenn de Comice - 3-4 months

Ideal demands:

0°C (32°F), 90-95% RH

Recommended temperature:

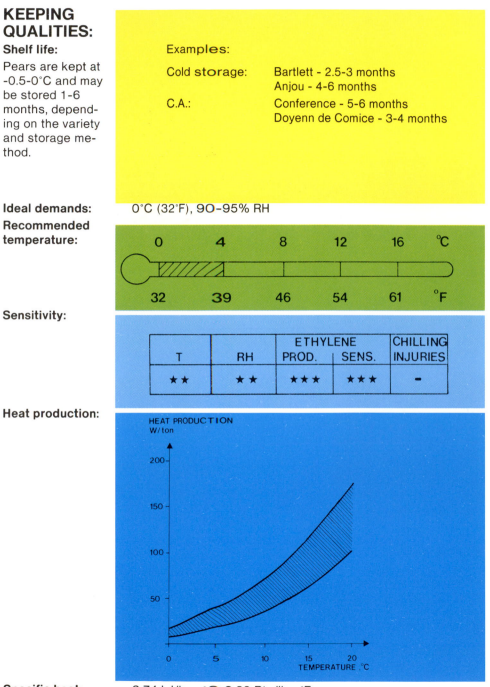

0	4	8	12	16	°C
32	39	46	54	61	°F

Sensitivity:

T	RH	ETHYLENE PROD.	SENS.	CHILLING INJURIES
★★	★★	★★★	★★★	-

Heat production:

HEAT PRODUCTION
W/ton

200

150

100

50

0   5   10   15   20
TEMPERATURE ,°C

Specific heat:

3.74 kJ/kg x °C   0.89 Btu/lb x °F

Specific weight:

Palletized boxes 350-400 kg/m³ Bulk. approx. 600 kg/m³

DESCRIPTION OF PRODUCT:

The pineapple plant resembles the agave which originates from Brazil. The plant has serrated leaves that are over a metre long. The fruit forms on a short, sturdy stem that shoots up from the middle of the plant. The stem grows through the fruit and ends in a tuft of leaves. Pineapples are cultivated throughout the tropical regions.

greenish-yellow or golden. Pineapples should be free from disease, rot or mould and have no foreign smell or taste. The flesh of the fruit should be golden, juicy and firm without being fibrous. Pineapples should have no chilling injuries which may result in brown or dull black skin, watery flesh and wilted leaves.

MAJOR PRODUCERS:

The major producers of pineapples are Thailand, the Philippines, Brazil, India and Vietnam.

STANDARDS:

There are no international standards for pineapples and although there are many recommendations, e.g. U.S. Grade Standards, they are not mandatory in international trade.

MINIMUM REQUIREMENTS:

Pineapples must be fresh and firm. The top shoot should be green and undamaged. The surface of the fruit should be

Latin:	Ananas comosus (L) Merr.
French:	Ananas
German:	Ananas
Spanish:	Ananás or Pira de America

PINEAPPLE

KEEPING QUALITIES:

Shelf life:

Storage temperature depends on the stage of ripeness.

Unripe:	10°C, 2-3 weeks
Ripe:	7-8°C, 5-7 days
	20°C, approx. 3 days

Ideal demands: 10°C (50°F), 90-95% RH

Recommended temperature:

| 0 | 4 | 8 | 12 | 16 | °C |
| 32 | 39 | 46 | 54 | 61 | °F |

Sensitivity:

T	RH	ETHYLENE PROD.	SENS.	CHILLING INJURIES
★★	★	★	–	★★★

Heat production:

HEAT PRODUCTION
W/ton

200

150

100

50

5   10   15   20   25

TEMPERATURE, °C

Specific heat: 3.68 kJ/kg x °C 0.88 Btu/lb x °F
Specific weight: Palletized cartons 200-250 kg/m³

DESCRIPTION OF PRODUCT:

Plums are believed to have come into existence in Asia Minor as a hybrid of cherry and sloe. The plum tree is a medium-sized tree (about 6 m in height). The fruit has a thin and often tough skin covered with a waxy bloom which reduces evaporation and contains a stone.

There are many varieties of plum in various colours, shapes and sizes. Ripe plums are soft and bruise easily. A large proportion of the plum crop is either dried or processed.

MAJOR PRODUCERS:

The major producers of plums are U.S.S.R, Rumania, China, Yugoslavia and West Germany.

STANDARDS:

For trade within the EEC plums must fulfil EEC Standard No. 6c. Additionally, there are many recommendations, e.g. U.S. Grade Standards, but they are not mandatory in international trade.

MINIMUM REQUIREMENTS:

Plums should be intact, clean, sound and free of foreign smell or taste. They should be firm and free of cracked skin and flesh, bruises and blemishes. The bloom should practically cover most of the fruit. They should bear no signs of attack by disease, pests, mould or rot.

Latin: Prunus domestica L.
French: Prune (commune)
German: Pflaume
Spanish: Ciruela

PLUM

KEEPING QUALITIES:
Shelf life:

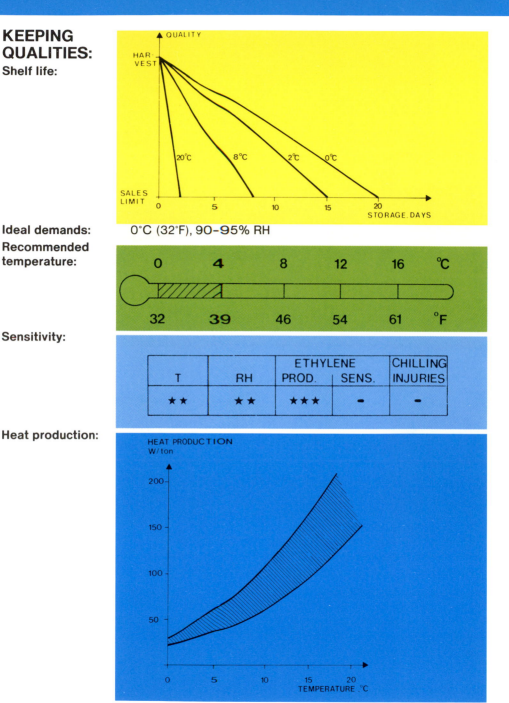

QUALITY

HAR-VEST

20°C   8°C   2°C   0°C

SALES LIMIT

0   5   10   15   20

STORAGE. DAYS

Ideal demands: 0°C (32°F), 90–95% RH

Recommended temperature:

| 0 | 4 | 8 | 12 | 16 | °C |
| 32 | 39 | 46 | 54 | 61 | °F |

Sensitivity:

T	RH	ETHYLENE PROD.	SENS.	CHILLING INJURIES
★★	★★	★★★	-	-

Heat production:

HEAT PRODUCTION W/ton

200

150

100

50

0   5   10   15   20

TEMPERATURE .°C

Specific heat: 3.68 kJ/kg x °C 0.88 Btu/lb x °F
Specific weight: Palletized boxes 350-400 kg/m³. Bulk approx. 600 kg/m³

195

POMEGRANATE

Latin: Punica granatum L.
French: Grenade
German: Granatapfel
Spanish: Granada

DESCRIPTION OF PRODUCT:

The pomegranate probably has its origin in western Asia, but grows wild in many places. It has been cultivated for thousands of years in the Middle East and brought to Spain where the town Granada, which is still an important area for the cultivation of pomegranates, was named after it.

The plant is a bush or a small tree. A pomegranate is the size of an apple (6-8 cm in diameter). The hard, leathery rind is reddish or brown and bears a protruding remnant of the calyx.

The fruit is full of tiny white kernels which are enclosed in juicy red pulp within pale walled sections. The pulp is the edible part of the fruit. Pomegranates have a sweet-sour taste.

MAJOR PRODUCERS:

The major producers of pomegranates are Spain, Italy, Israel, South Africa and U.S.A.

STANDARDS:

There are no international standards for pomegranates.

MINIMUM REQUIREMENTS:

Pomegranates should be free from disease and pests, spots damage and cracks, and the rind should not be dried out. The edible part of the fruit should be juicy and free from foreign smell or taste.

196

KEEPING QUALITIES:

Shelf life:

0°C, 90% RH, 2 months
20°C, 60% RH, 1-2 weeks

Ideal demands:

0-2°C (32-36°F), 90-95% RH

Recommended temperature:

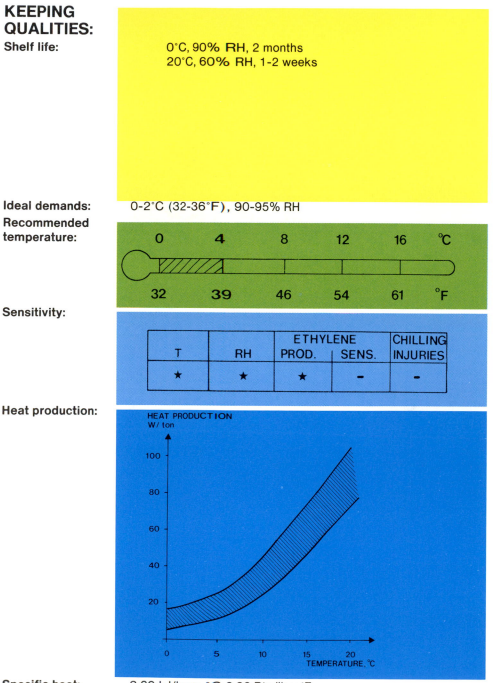

| 0 | 4 | 8 | 12 | 16 | °C |

| 32 | 39 | 46 | 54 | 61 | °F |

Sensitivity:

T	RH	ETHYLENE PROD.	SENS.	CHILLING INJURIES
★	★	★	–	–

Heat production:

HEAT PRODUCTION
W/ ton

TEMPERATURE, °C

Specific heat: 3.60 kJ/kg x °C 0.86 Btu/lb x °F

Specific weight: Palletized cartons 300-350 kg/m³. Bulk approx. 500 kg/m³

197

DESCRIPTION OF PRODUCT:

The potato is native to northern South America where it has been cultivated for several thousand years.

It is a short-day plant and demands a specific climate for growth. It is an herbaceous plant approx. 0.5 m tall. The edible tubers are brown with a thin corky skin that protects against moisture loss and disease. Exposure to sunlight causes the tubers to turn green producing a poisonous alkaloid, solanine, in the tissues. Low temperatures produce a sweet taste caused by hydrolysis of starch into sugar compounds.

MAJOR PRODUCERS:

The major producers of potatoes are U.S.S.R., China, Poland, U.S.A. and India.

STANDARDS:

There are no international standards for potatoes. However, there are many recommendations, e.g. ECE Standard No. FFV-31 and U.S. Grade Standards, but they are not mandatory in international trade.

MINIMUM REQUIREMENTS:

Potatoes should be intact, sound, clean, free from soil and other extraneous matter, firm and should not emit any foreign smell or taste. They should not be affected by greening or sprouting nor bear signs of disease, rot, mould or pests. They should be free from mechanical damage, bruises or other defects, external as well as internal.

Latin:	Solanum tuberosum L.
French:	Pomme de terre
German:	Kartoffel
Spanish:	Patata

POTATO

KEEPING QUALITIES:

Shelf life:

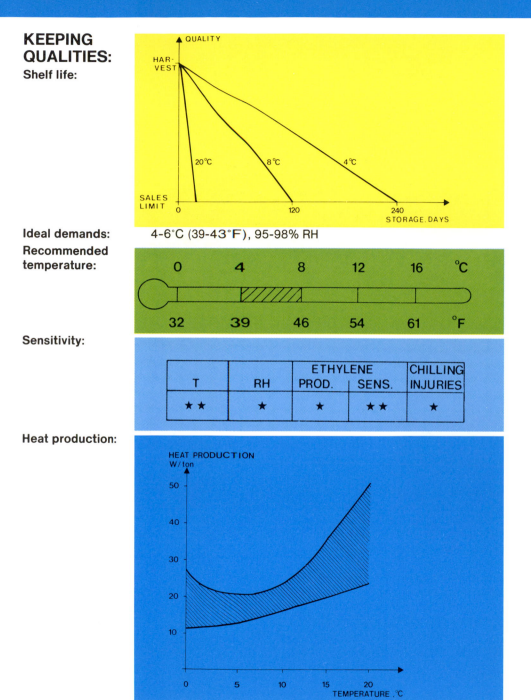

Ideal demands: 4-6°C (39-43°F), 95-98% RH

Recommended temperature:

Sensitivity:

T	RH	ETHYLENE PROD.	SENS.	CHILLING INJURIES
★ ★	★	★	★ ★	★

Heat production:

Specific heat: 3.60 kJ/kg x °C 0.86 Btu/lb x °F

Specific weight: Palletized crates 330-350 kg/m³. Bulk approx. 650 kg/m³

(Cactus fruit or Indian fig)

PRICKLY PEAR

Latin: Opuntia ficus-indica (L.) Miller
French: Figne d'Inde
German: Kaktusfeige or Opuntie
Spanish: Higo chumbo

DESCRIPTION OF PRODUCT:

Prickly pears have their origin in Mexico where they still play an important role in arid regions. The plant is a 4 m tall cactus, whose stem comprises flat disc-shaped links. The flowers are small and yellow.

The fruit resembles the pear in shape and size. The skin is green when unripe, becoming yellow, green, red or orange on ripening. The flesh is yellowish orange or red and contains a number of small seeds that may also be consumed. Prickly pears are very juicy and their consistency resembles watermelon. They have a sweet, but not particularly pronounced taste.

There are fine, small, almost invisible hairs on the skin which, when touched, penetrate the skin. Gloves should therefore be used when handling prickly pears.

MAJOR PRODUCERS:

The major producers of prickly pears are Mexico, Italy, Spain, Israel and Brazil.

STANDARDS:

There are no international standards for prickly pears.

MINIMUM REQUIREMENTS:

Prickly pears should be intact, clean and sound. The fruit should be free from visible attack by disease or insects and from foreign smell or taste. The skin should be undamaged and free from bruises and pressure marks. The colour should be an even red, yellow, orange or green depending on the variety. The ripe fruit is soft.

Latin: Opuntia ficus-indica (L.) Miller
French: Figne d'Inde
German: Kaktusfeige or Opuntie
Spanish: Higo chumbo

(Cactus fruit or Indian fig)
PRICKLY PEAR

KEEPING QUALITIES:

Shelf life:

5°C, 90% RH, 3-4 weeks
20°C, 60% RH, 5-7 days

Ideal demands:

5°C (41°F), 90–95% RH

Recommended temperature:

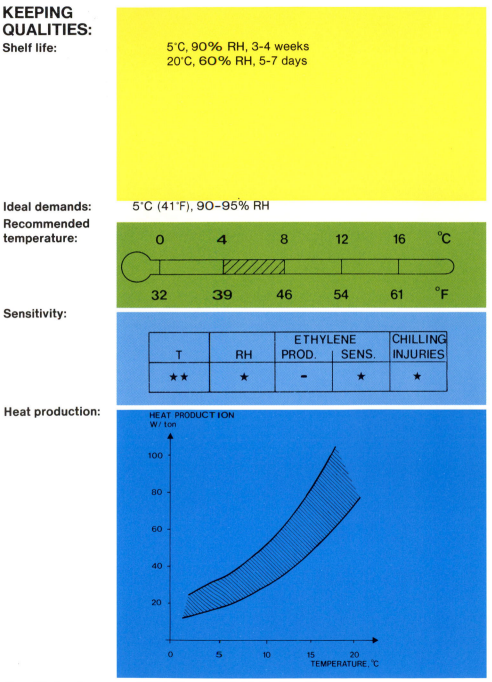

Sensitivity:

T	RH	ETHYLENE PROD.	SENS.	CHILLING INJURIES
★★	★	–	★	★

Heat production:

HEAT PRODUCTION
W/ton

Specific heat: 3.72 kJ/kg x °C 0.89 Btu/lb x °F
Specific weight: Boxes, pallets 280-350 kg/m³. Bulk approx. 550 kg/m³

DESCRIPTION OF PRODUCT:

There are two species of pumpkin -
cucurbita maxima, which is round or flatly round, and cucurbita pepo, which is
oblong. Both species come from Latin
America where they were cultivated by
the Indians.

The plant, a coarse trailing vine, is an
annual herb. The fruit consists of a thick
rind enclosing the pulp with a spongy, web-like core containing numerous
seeds.

There is a large number of varieties. In
breeding, selection criteria favour a fruit
with thick flesh and a small core.

Pumpkins can be very large and a weight
of 50 kg is not unusual.

MAJOR PRODUCERS:

The major producers of pumpkins are China, Romania, Egypt, Argentine and Turkey.

STANDARDS:

There are no international standards for
pumpkins.

MINIMUM REQUIREMENTS:

Pumpkins should be intact, clean, sound
and fresh in appearance. They should
not be infected by disease or insects.
They should be free from bruises caused
by handling or any other type of damage. The flesh should be pale, without fibres, water-logged areas or woodiness.

Latin: Curcurbita spp.
French: Potiron
German: Kürbis
Spanish: Calabaza

Winter squash) **PUMPKIN**

KEEPING QUALITIES:

Shelf life:

10°C, 75% RH, 2-3 months
20°C, 60% RH, 2-4 weeks

Ideal demands: 7-10°C (45-50°F), 75-85% RH

Recommended temperature:

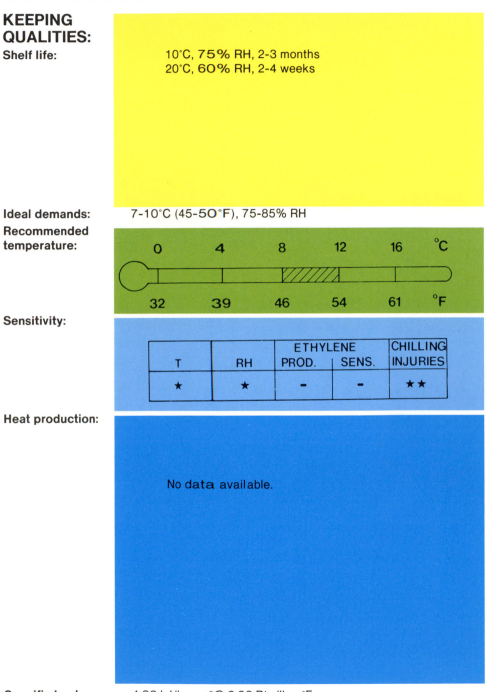

| | | ETHYLENE | | CHILLING |
T	RH	PROD.	SENS.	INJURIES
★	★	–	–	★★

Sensitivity:

Heat production:

No data available.

Specific heat: 4.02 kJ/kg x °C 0.96 Btu/lb x °F

Specific weight: Palletized boxes 200-300 kg/m³. Bulk approx. 500 kg/m³

DESCRIPTION OF PRODUCT:

Radish is one of the oldest cultivated plants and it is believed to originate from China where it has been known for thousands of years. Egyptians knew of the radish before the pyramids were built, whilst the Romans learned of it around the beginning of the Christian era.

The plant is an annual. The edible part of the radish is formed by the swelling of the hypocotyl, which means that botanically the tuber is part of the stem as well as the root. Radishes may vary considerably in shape and size, from round to conical and from red to red–white and white.

The pungent taste comes from its content of mustard oils. These are largely concentrated in the outer layer - the content may be 2-3 times higher than in the middle of the radish.

MAJOR PRODUCERS:

The major producers of radish are France, Italy, Holland, Spain and U.S.A.

STANDARDS:

There are no international standards for radishes. However, there are many recommendations, e.g. OECD Standards and U.S. Grade Standards, but they are not mandatory in international trade.

MINIMUM REQUIREMENTS:

Radishes should be clean, sound and intact. They should be fully developed and of a colour and shape typical of the variety. Radishes should be free from damage, rust or signs of attack by insects, disease, rot or mould. They should not be hollow or woody. There should be no foreign smell or taste. Radishes in a batch should be of the same variety. If radishes are sold with leaves these should be fresh, and green. Otherwise, the radishes should have a minimum of inedible parts.

KEEPING QUALITIES:

Shelf life:

Radishes with tops:
- 0-1°C, 3-5 days
- 2-5°C, 2-3 days
- 20°C, 1 day

Radishes without tops:
- 0-1°C, 10-14 days
- 2-5°C, 7-10 days
- 20°C, 2 days

Ideal demands: 0°C (32°F), 90-95% RH

Recommended temperature:

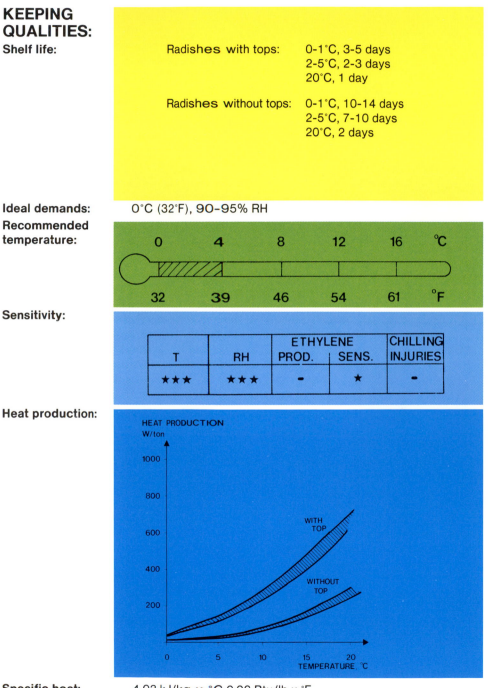

Sensitivity:

T	RH	ETHYLENE PROD.	SENS.	CHILLING INJURIES
★★★	★★★	-	★	-

Heat production:

Specific heat: 4.03 kJ/kg x °C 0.96 Btu/lb x °F

Specific weight: Boxes, pallets 230-300 kg/m³. Bulk approx. 550 kg/m³

RASPBERRY

Latin: Rubus idaeus L.
French: Framboise
German: Himbeere
Spanish: Frambuesa

DESCRIPTION OF PRODUCT:

Wild varieties of raspberries are found in Europe, West Asia and North America and its origin is, therefore, uncertain. The picking of wild raspberries is still quite common.

Cultivation of raspberries began in the last century, starting in U.S.A. and later in Europe. Only red raspberries are cultivated in Europe, whereas in America black raspberries are also grown.

The plant is approx. 1 m tall - a shrub with thorny shoots that live 1-2 years. The berries are about 2 cm long, cup-shaped and comprise a number of fleshy and juicy small sections - drupelets - each with a seed inside.

MAJOR PRODUCERS:

The major producers of raspberries are U.S.S.R., Poland, Yugoslavia, Hungary and West Germany.

STANDARDS:

There are no international standards for raspberries. On the other hand, there are several recommendations, e.g. ECE Standard No. FFV-32 and U.S. Grade Standards, but they are not mandatory in international trade.

MINIMUM REQUIREMENTS:

Raspberries should be clean, sound, firm and juicy without wilting or shrivelling. They should be ripe, but not overripe, and taste characteristically of raspberry with no foreign taste or smell. Raspberries should be fully developed without underdeveloped drupelets or sections. They should be uninfected by pests, mould, rot or diseases.

KEEPING QUALITIES:

Shelf life:

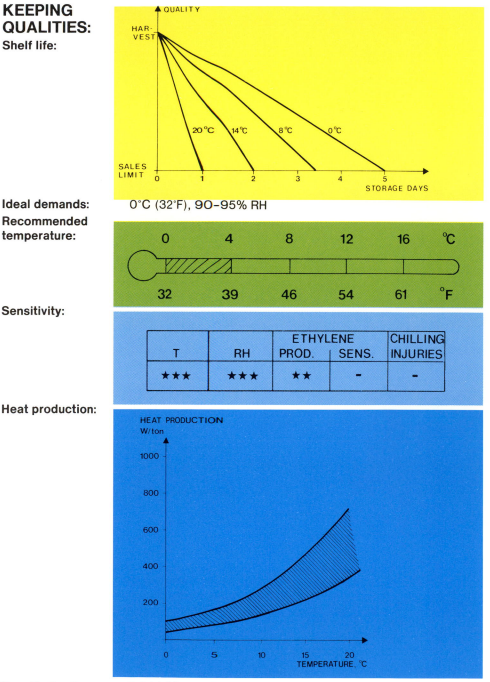

Ideal demands: 0°C (32°F), 90–95% RH

Recommended temperature:

Sensitivity:

T	RH	ETHYLENE PROD.	SENS.	CHILLING INJURIES
★★★	★★★	★★	-	-

Heat production:

Specific heat: 3.64 kJ/kg x °C 0.87 Btu/lb x °F

Specific weight: Palletized cartons 250-350 kg/m³. Bulk approx. 650 kg/m³

207

RED BEET (Beetroot)

DESCRIPTION OF PRODUCT:

It is believed that red beet (sugar beet, leafy beet or Swiss chard, etc.) has been derived from sea beet. The cultivation of beets has been known for more than 3000 years, but originally it was only the leaves and the stems that were used. Only from the 2nd and 3rd century did the use of roots for human consumption become known. Red beet is a relatively new plant – from around the Middle Ages.

This plant is a biennial. It forms a rosette of leaves during the first year and a root which stores the plant's nutrients. The beet itself does not only comprise the root, but also a part of the attached stem.

There are various shapes of beetroot ranging from round to cylindrical and oblong (the picture above shows the 2 first-named shapes). The round and the cylindrical types are mainly used as fresh produce, whilst the oblong is usually processed.

MAJOR PRODUCERS:

The major producers of beetroot are West Germany, the Netherlands, Poland, U.S.A. and U.S.S.R.

STANDARDS:

There are no international standards for beetroots. Many recommendations are to be found, e.g. U.S. Grade Standards, but they are not mandatory in international trade.

MINIMUM REQUIREMENTS:

Beetroots should be intact, clean, sound, without deformities and any foreign smell or taste. They should be firm and full-bodied. The tops should be removed without damaging the root. The roots should only bear few signs of attack by insects or disease and have no shoots and woodiness. The shape should be typical of the variety and different shapes should not be mixed in a batch. The roots should be even in size.

Latin: Beta vulgaris L. var. conditiva **Alef.**
French: Betterave rouge
German: Rote Rübe or Rote Bete
Spanish: Remolacha hortelana

(Beetroot) **RED BEET**

KEEPING QUALITIES:

Shelf life:

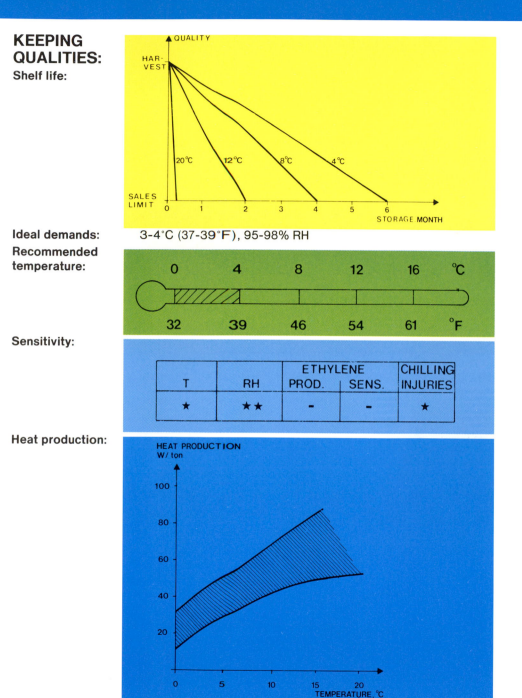

Ideal demands: 3-4°C (37-39°F), 95-98% RH

Recommended temperature:

0	4	8	12	16	°C
32	39	46	54	61	°F

Sensitivity:

T	RH	ETHYLENE PROD.	SENS.	CHILLING INJURIES
★	★★	-	-	★

Heat production:

Specific heat: 3.92 kJ/kg x °C 0.94 Btu/lb x °F
Specific weight: Boxes, pallets 350-400 kg/m³. Bulk approx. 600 kg/m³

Latin:	Vaccinium vitis idaea L.
French:	Airelle rouge
German:	Preiselbeere
Spanish:	Arendano rojo

DESCRIPTION OF PRODUCT:

Red whortleberries originate from northern and central Europe where they are found growing wild in many places. This shrub grows in damp, acid ground and can also be found on heaths and in pine forests. It has been known for several thousand years. Actual cultivation of this shrub is insignificant. Most of the berries that are on the market are gathered from the wild.

The red whortleberry itself is very decorative. The shrub has leathery dark green leaves which are not shed in winter. It is most beautiful in September when it is covered with shiny, red berries which grow in small clusters at the tip of the shoots.

The fresh berries are rather bitter and cannot be eaten raw. The berries are juicy and 5–10 mm in diameter. They contain a considerable amount of benzoic acid.

MAJOR PRODUCERS:

The major producers of red whortleberries are West Germany, Poland, Austria and Scandinavia.

STANDARDS:

There are no international standards for red whortleberries.

MINIMUM REQUIREMENTS:

Red whortleberries should be fresh, intact and clean. They should have a characteristic taste and be free from foreign smell or taste and signs of attack by disease and pests. The berries should be full-bodied and adequately ripe. Dried out, mechanically damaged, bruised and cracked berries should not be included. Leaves, stalks and other inedible parts of the plant should be removed.

Latin:	Vaccinium vitis idaea L.
French:	Airelle rouge
German:	Preiselbeere
Spanish:	Arendano rojo

(Cowberry)
RED **WHORTLEBERRY**

KEEPING QUALITIES:

Shelf life:

0°C, **90%** RH, 3-4 weeks
20°C, **60%** RH, 5-7 days

Ideal demands: 0°C (32°F), 90–95% RH

Recommended temperature:

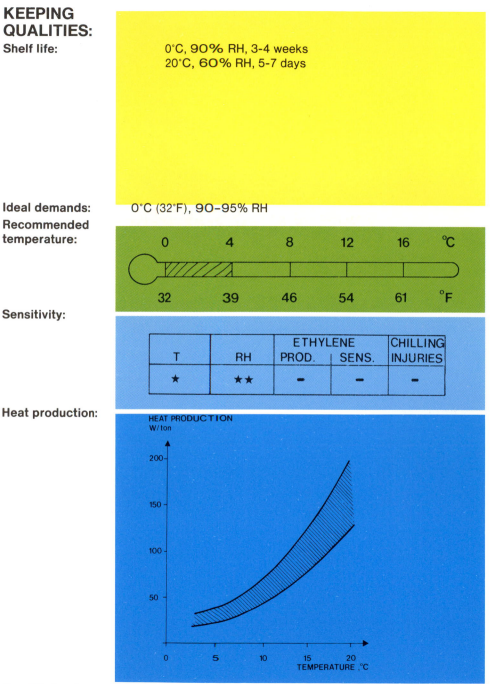

Sensitivity:

T	RH	ETHYLENE PROD.	SENS.	CHILLING INJURIES
★	★★	–	–	–

Heat production:

HEAT PRODUCTION
W/ton

Specific heat: 3.68 kJ/kg x °C 0.88 Btu/lb x °F
Specific weight: Boxes, pallets 300-350 kg/m³. Bulk approx. 600 kg/m³

DESCRIPTION OF PRODUCT:

Rhubarb came originally from the belt that stretches across the south of Siberia along the River Volga and has been known in Tibet and Mongolia for several thousand years. It has been cultivated in Italy and England for approx. 400 years, but only much later did it become commonly known. The name rhabarbarum, as the Romans called it, means »barbarians' plant».

The rhubarb is a hardy perennial which may be harvested for 5-10 years. The inflorescence which may grow to a height of 2 m is branched and bears small cream coloured flowers.

The leaf stalks, the edible part of the plant, are juicy, green or red. Old stalks are fibrous. The normal length of a stalk is approx. 50 cm.

MAJOR PRODUCERS:

The major producers of rhubarb are U.K., West Germany, U.S.A. and the Netherlands.

STANDARDS:

There are no international standards for rhubarb. However, many recommendations are to be found, e.g. U.S. Grade Standards and OECD Standards, but they are not mandatory in international trade.

MINIMUM REQUIREMENTS:

Rhubarb should be intact, fresh, sound and clean. The leaves should be removed from the stalks. If the leaves are still attached they should be full-bodied and green. The root end of the stalks should also be removed. Rhubarb stalks should be free from any signs of attack by disease, insects, rot or mould. They should not be mechanically damaged or spotted. The stalks should not be fibrous.

KEEPING QUALITIES:
Shelf life:

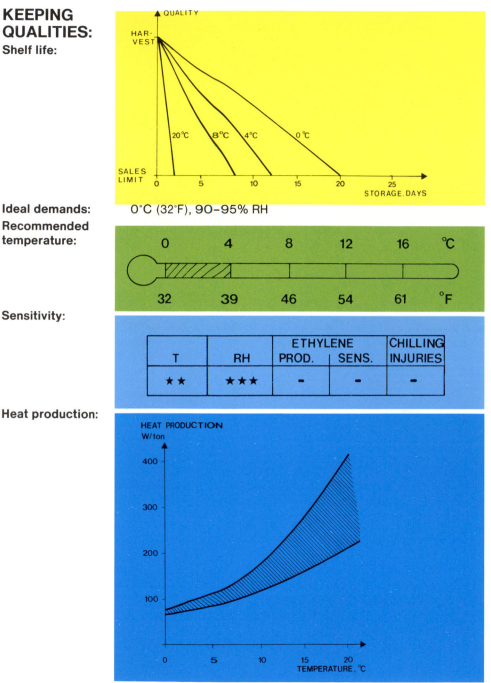

Ideal demands: 0°C (32°F), 90–95% RH

Recommended temperature:

0	4	8	12	16	°C
32	39	46	54	61	°F

Sensitivity:

T	RH	ETHYLENE PROD.	SENS.	CHILLING INJURIES
★★	★★★	–	–	–

Heat production:

Specific heat: 4.04 kJ/kg x °C 0.97 Btu/lb x °F
Specific weight: Boxes, pallets 250–300 kg/m³. Bulk approx. 750 kg/m³

CABBAGE, SAVOY

Latin:	Brassica oleracea L. convar. capitata (L.) Alef. var. sabauda L.
French:	Chou de Milan
German:	Wirsingkohl
Spanish:	Col de Milán

DESCRIPTION OF PRODUCT:

The cultivated cabbage originates from wild species which originally grew in Asia Minor. From there they spread to the coastal regions around the Mediterranean and along the Atlantic Ocean. The earliest varieties were single-leafed. In the year 800 the first headed cabbages began to appear, but they had much looser heads than those found today.

Savoy and other cabbages share an essentially common history. It is believed to have come into existence as an independent variety in the Middle Ages in Savoy, south-eastern France.

Savoy cabbage is recognizable by its wavy, wrinkled leaves. This type of cabbage has a milder taste. There are both yellow and green types; the green is most common. Savoy cabbage is more winter hardy than white cabbage, which means that it can be harvested later than white cabbage. In mild regions it may be left in the fields throughout the winter.

MAJOR PRODUCERS:

The major producers of Savoy cabbage are West Germany, France, U.K., Poland and U.S.S.R.

STANDARDS:

For trade within the EEC Savoy cabbage must comply with EEC Standard No. 24. There are many recommendations, e.g. U.S. Grade Standards, but they are not mandatory in international trade.

MINIMUM REQUIREMENTS:

Savoy cabbage should be full-bodied, fresh and of a colour typical of the variety. The cabbage should be intact with a firm head which is free from soil, insects, rot and mould. The stem should be cut directly below the lowest leaf. Savoy cabbage should be free from frost damage, spots and mechanical damage and from foreign smell or taste.

Latin: Brassica oleracea L. convar. capitata (L.)
Alef. var. sabauda L.
French: Chou de Milan
German: Wirsingkohl
Spanish: Col de Milán

SAVOY **CABBAGE**

KEEPING QUALITIES:
Shelf life:

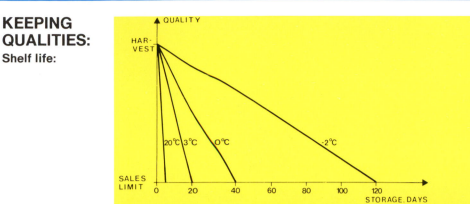

Ideal demands: -2-0°C (28-32°F), 90-95% RH

Recommended temperature:

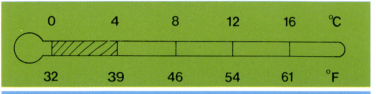

Sensitivity:

T	RH	ETHYLENE PROD.	SENS.	CHILLING INJURIES
★ ★	★ ★	★	★ ★	–

Heat production:

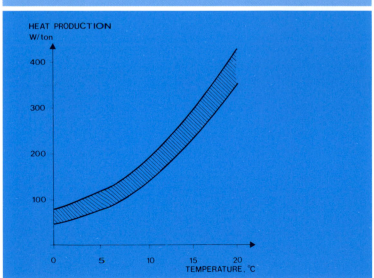

Specific heat: 3.93 kJ/kg x °C 0.94 Btu/lb x °F
Specific weight: Boxes, pallets 200-300 kg/m³. Bulk approx. 550 kg/m³

215

DESCRIPTION OF PRODUCT:

The name is believed to have come from the Italian word, scorzone, which means poisonous, black snake. It is presumed to have its origin in Spain, hence the name hispanica. This vegetable has been cultivated especially in France and Belgium where it is still popular.

Scorzonera has a long white-fleshed root (up to 30 cm long) with a black-brown bark. The leaves are long and slender and may be used in salads. The plant can easily survive the winter and flowers in its second year, but unlike other vegetables the root can still be used.

The root does not contain starch as nutrients, but insulin which is a carbohydrate mainly comprising units of fructose and only a few percent glucose. This makes it of interest to diabetics. Scorzonera provides energy and is rich in minerals, especially iron. The root has a white, milky juice which darkens on contact with air. This may be avoided by boiling the roots with the skin on.

MAJOR PRODUCERS:

The major producers of scorzonera are Belgium, the Netherlands and France.

STANDARDS:

There are no international standards for scorzonera. There are many recommendations, e.g. OECD Standards and ECE Standard No. FFV-33, but they are not mandatory in international trade.

MINIMUM REQUIREMENTS:

Scorzonera should be intact, clean, fresh and without any foreign smell or taste. The roots should be full-bodied, straight and unbranched. Scorzonera should not be woody. The leaves should be removed without damaging the root. There should be no signs of superficial damage, regrowth, rot, disease, mould or insects.

KEEPING QUALITIES:

Shelf life:

0°C, **90% RH**, 4 months
20°C, **60% RH**, 5-7 days

Ideal demands: 0°C (32°F), 90–95% RH

Recommended temperature:

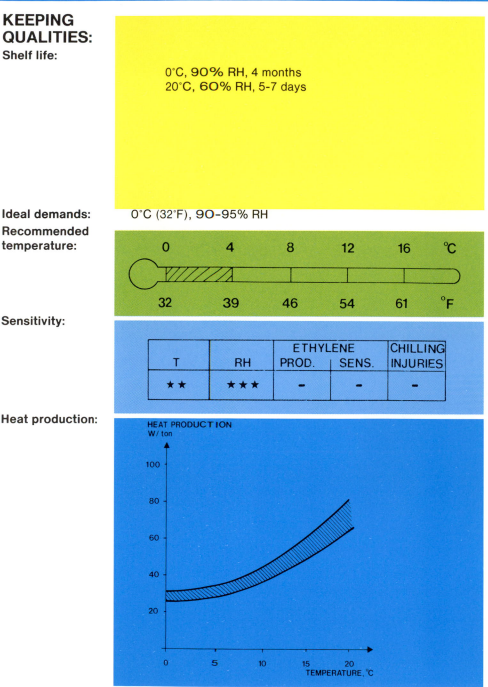

	T	RH	ETHYLENE PROD.	SENS.	CHILLING INJURIES
	★★	★★★	–	–	–

Sensitivity:

Heat production:

Specific heat: 3.50 kJ/kg x °C 0.83 Btu/lb x °F
Specific weight: Boxes, pallets 250-350 kg/m³. Bulk approx. 500 kg/m³

(Bitter orange or sour orange)

ORANGE, SEVILLE

Latin: Citrus aurantium L.
French: Bigaradier
German: Pomeranze or Bitterorange
Spanish: Naranjo amargo or naranjo agrio

DESCRIPTION OF PRODUCT:

Seville orange is native to India. It was brought by the Arabs to Seville which is noted for its vast production of bitter oranges. It is now cultivated in many subtropical regions.

It is generally believed that bitter orange may be the origin of all our well-known citrus forms. Bitter orange is a tall tree, up to 15 m, with thin thorns and dark green foliage. The flowers are axillary, large, white and very fragrant.

Bitter orange rind is used in the production of liqueur and the flowers in perfumes.

The fruit is dark orange, almost reddish. The peel is very thick with large pores and the flesh is quite dry.

The very spicy fragrance of bitter oranges makes them easily distinguishable from sweet oranges which they closely resemble. As the name suggests, bitter oranges are bitter and sour, but very aromatic.

MAJOR PRODUCERS:

The major producers of bitter oranges are India, Spain, Italy and Latin America.

STANDARDS:

There are no international standards for bitter oranges.

MINIMUM REQUIREMENTS:

Bitter oranges should be intact, sound, free from frost damage and visible foreign matter with no foreign smell or taste. The fruits should be free of bruises, cracks, mechanical damage and spots. Bitter oranges should be full-bodied and firm. The colour should be even and typical of the variety.

Latin: Citrus aurantium L.
French: Bigaradier
German: Pomeranze or Bitterorange
Spanish: Naranjo amargo or naranjo agrio

(Bitter orange or sour orange)
SEVILLE **ORANGE**

KEEPING QUALITIES:

Shelf life:

10°C, 90% RH, 3 months
20°C, 60% RH, 2-3 weeks

Ideal demands: 10°C (50°F), 85-90% RH

Recommended temperature:

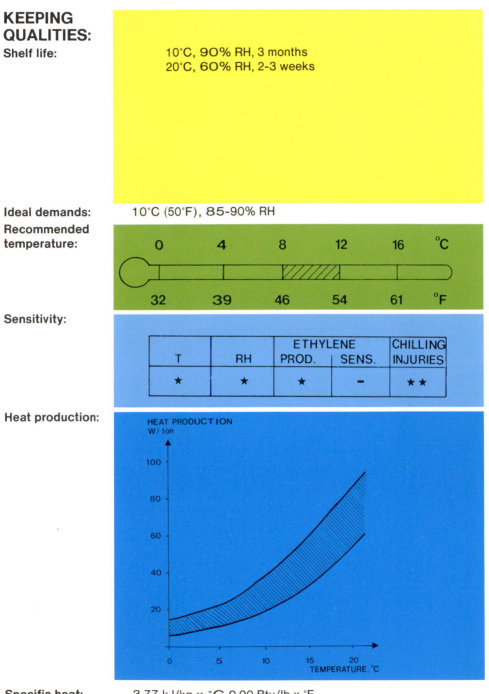

| | | ETHYLENE | | CHILLING |
T	RH	PROD.	SENS.	INJURIES
★	★	★	−	★★

Sensitivity:

Heat production:

HEAT PRODUCTION
W/ ton

TEMPERATURE, °C

Specific heat: 3.77 kJ/kg x °C 0.90 Btu/lb x °F
Specific weight: Boxes, pallets 350-450 kg/m³. Bulk 480-520 kg/m³

DESCRIPTION OF PRODUCT:

Spinach originates from Iran where it has been known for thousands of years, but knowledge of spinach cultivation spread to other countries only two thousand years ago. It was brought to China in the 7th century and in the 12th century it was introduced to the Spaniards.

Spinach is an annual non-headed leafy vegetable. The period of growth from sowing to harvest is short which means several crops of spinach may be grown in the same field during one season. In summer, spinach tends to run to seed easily promoted by long days and high temperature. Spinach may winter in the fields. It is sold as loose leaves or whole plants.

Spinach must not be mistaken for New Zealand spinach (Tetragonia tetragonio-des). As the name suggests it comes from New Zealand and is not related to spinach. It has spreading, 1-meter long stalks with three or four-sided fleshy leaves

The content of oxalic acid in spinach binds calcium in the system thus preventing its assimilation. This is a notable factor if spinach is consumed very often.

MAJOR PRODUCERS:

Spinach is cultivated all over the world. Europe, North America and western Asia are important producers.

STANDARDS:

For trade within the EEC spinach must comply with EEC Standard No. 13. There are many recommendations, e.g. U.S. Grade Standards for both spinach leaves and spinach plants, but they are not mandatory in international trade.

MINIMUM REQUIREMENTS:

Spinach should be fresh, sound, clean and free from signs of running to seed or attack by pests. It should be free of frost and mechanical damage and of foreign smell or taste. If sold as loose leaves, no roots should be included and stalks should be no longer than 10 cm. Whole plants should be cut just below the lowest leaf.

KEEPING QUALITIES:
Shelf life:

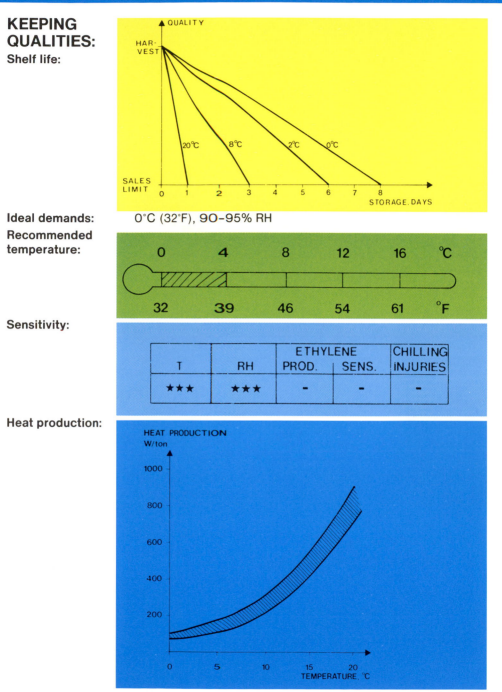

Ideal demands: 0°C (32°F), 90–95% RH

Recommended temperature:

					°C
0	4	8	12	16	
32	39	46	54	61	°F

Sensitivity:

T	RH	ETHYLENE PROD.	SENS.	CHILLING INJURIES
★★★	★★★	–	–	–

Heat production:

Specific heat: 3.93 kJ/kg x °C 0.94 Btu/lb x °F

Specific weight: Boxes, pallets 130-150 kg/m³. Bulk approx. 200 kg/m³

221

STRAWBERRY

Latin: Fragaria X ananassa Duch.
French: Fraise
German: Erdbeere
Spanish: Fresa

DESCRIPTION OF PRODUCT:

The origin of strawberries is unknown, as wild varieties are found in almost all parts of the world. As there is a large number of varieties, strawberries may be cultivated under various climatic conditions.

The plant is a perennial and does not require pollination. The fruit, or strawberry, is not a true fruit in the botanical sense, since it consists of the swollen receptacle on the surface of which are embedded the numerous achenes or seeds (the actual fruit).

MAJOR PRODUCERS:

The major producers of strawberries are U.S.A., Poland, Spain, Japan and Italy.

STANDARDS:

For trade within the EEC strawberries must comply with EEC Standard No. 11. There are many recommendations, e.g. U.S. Grade Standards, but they are not mandatory in international trade.

MINIMUM REQUIREMENTS:

Strawberries should be fresh, intact, clean and sound. Each berry should bear its green calyx (the hull). The berries should not be infected with disease, rot, mould or pests and be free from any foreign smell or taste. Strawberries should be adequately ripe and show no signs of bruising or damage.

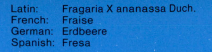

Latin: Fragaria X ananassa Duch.
French: Fraise
German: Erdbeere
Spanish: Fresa

STRAWBERRY

KEEPING QUALITIES:

Shelf life:

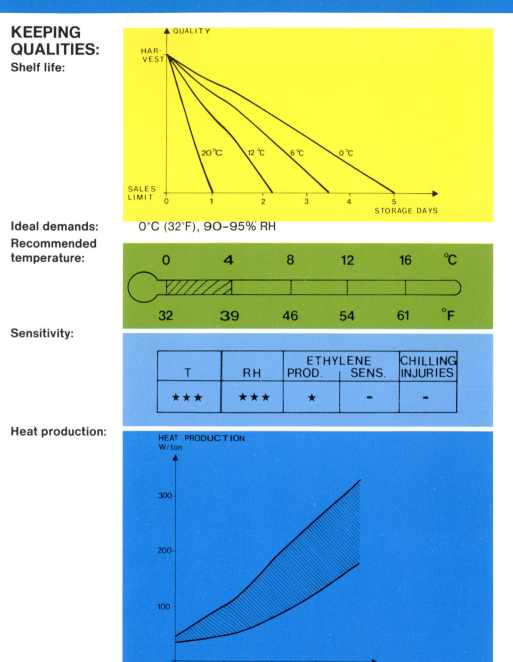

Ideal demands: 0°C (32°F), 90–95% RH

Recommended temperature:

Sensitivity:

	T	RH	ETHYLENE PROD.	ETHYLENE SENS.	CHILLING INJURIES
	★★★	★★★	★	-	-

Heat production:

Specific heat: 3.85 kJ/kg x °C 0.92 Btu/lb x °F

Specific weight: Palletized boxes 160-200 kg/m³. Bulk approx. 550 kg/m³

223

SUGAR PEA (Mange-tout pea)

Latin:	Pisum sativum L.
French:	Pois mange tout or pois sucré
German:	Zuckererbse
Spanish:	Guisante azucarado or Tirabeque

DESCRIPTION OF PRODUCT:

Peas originate from the eastern Mediterranean region, Iran, Afghanistan and Tibet. They have been known and cultivated for thousands of years.

The plant is an annual twining herb which usually has a long stalk of several metres with branching tendrils.

The most common peas for fresh consumption are marrowfat which have large, soft seeds and a tough pod. The sugar pea pod does not have this tough membrane and is, therefore, edible.

Sugar pea is harvested when the pod is flat, before the seeds begin to develop. The length of the pods is typically 7-11 cm.

MAJOR PRODUCERS:

The major producers of peas, including the sugar pea, are U.S.A., U.K., France, Hungary and China.

STANDARDS:

There are no international standards for sugar pea.

MINIMUM REQUIREMENTS:

Sugar peas should have intact, clean, juicy and sound pods. The seeds should be undeveloped and this should be clearly visible. The pods should not have a tough membrane and bear no signs of attack by disease, insects, rot and mould. Sugar peas should be evenly green and undamaged.

Latin: Pisum sativum L.
French: Pois mange tout or pois sucré
German: Zuckererbse
Spanish: Guisante azucarado or Tirabeque

(Mange-tout pea) **SUGAR PEA**

KEEPING QUALITIES:

Shelf life:

0°C, 90% RH, 1-2 weeks
20°C, 60% RH, 1-2 days

Ideal demands:

0°C (32°F), 90–95% RH

Recommended temperature:

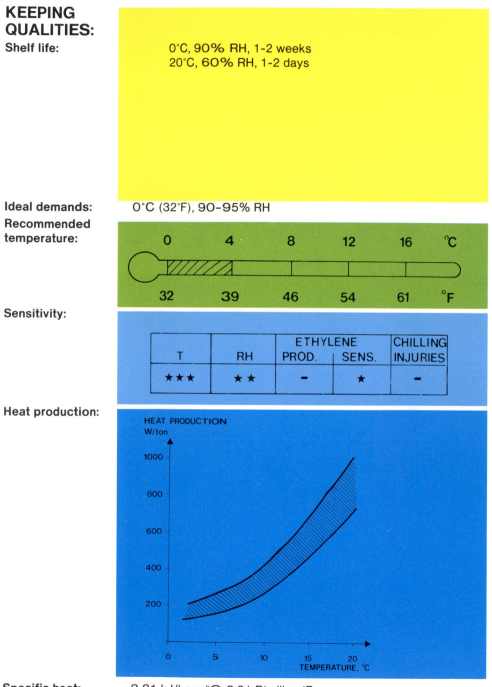

	T	RH	ETHYLENE PROD.	SENS.	CHILLING INJURIES
Sensitivity:	★★★	★★	–	★	–

Heat production:

HEAT PRODUCTION
W/ton

Specific heat: 3.81 kJ/kg x °C 0.91 Btu/lb x °F
Specific weight: Boxes, pallets 1 40-200 kg/m³

225

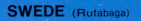

SWEDE (Rutabaga)

Latin:	Brassica napus L. var. napobrassica (L). Rchb.
French:	Chou navet or Rutabaga
German:	Kohlrübe or Steckrübe
Spanish:	Colirrabo

DESCRIPTION OF PRODUCT:

The swede, or rutabaga, is considered to be developed from rape, but there is some difference of opinion as to whether it originally came from Siberia or northern Europe. It is a very old cultivated plant.

The swede is used as food and livestock feed. It was and still is very popular in north European countries, especially in times of food shortage. As a forage plant it was most popular in the period from the First World War to the mid-Sixties.

This turnip is very nutritious. It is green or green-violet on the upper part and the flesh is white or yellow. Because of its high Vitamin C content it has been nicknamed "the lemon of the North". Swedes can become very large, but the small or medium-sized are preferred.

MAJOR PRODUCERS:

The major producers of swedes are the Scandinavian countries, West Germany, U.S.A., Canada and the Netherlands.

STANDARDS:

There are no international standards for swedes. Many recommendations are to be found, e.g. U.S. Grade Standards, but they are not mandatory in international trade.

MINIMUM REQUIREMENTS:

Swedes should be clean, intact and sound. Tops should be removed without damaging the roots. Shape, development and colour should be typical of the variety. Swedes should not be misshapen and they should be free from superficial damage, rifts or wounds. Swedes should not be affected by no foreign smell or taste. The flesh should be pale and free from any discolouration.

Latin: Brassica napus L. var. napobrassica (L). Rchb.
French: Chou navet or Rutabaga
German: Kohlrübe or Steckrübe
Spanish: Colirrabo

(Rutabaga) **SWEDE**

KEEPING QUALITIES:

Shelf life:

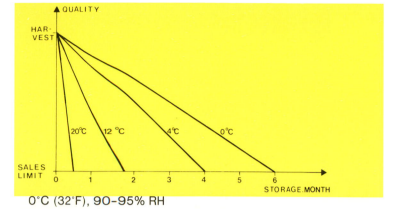

Ideal demands: 0°C (32°F), 90–95% RH

Recommended temperature:

0	4	8	12	16	°C
32	39	46	54	61	°F

Sensitivity:

T	RH	ETHYLENE PROD.	SENS.	CHILLING INJURIES
★	★	–	★	–

Heat production:

No data available.

Specific heat: 3.89 kJ/kg x °C 0.93 Btu/lb x °F

Specific weight: Boxes, pallets 200-300 kg/m^3. Bulk approx. 550 kg/m^3

SWEET CORN

Latin:	Zea mays L. var. saccharata Koern.
French:	Maïs doux
German:	Mais
Spanish:	Maiz

DESCRIPTION OF PRODUCT:

Sweet corn came originally from Latin America, where it was the staple crop of the Indians. It was mainly used in its mature form as a grain product, but the use of immature cobs was also known. At some time sweet corn came into existence, when or how is unknown. The main difference between these two sorts is that the process of converting sugar to starch in sweet corn is slower, which means that the kernels do not become mealy so quickly.

The maize plant has separate male and female flowers. Male flowers grow at the top of the main stalk, whilst the female flowers, which later become cobs, develop on side shoots. The »beard« on top of the cob is the stigma. Each thread leads to a kernel and all threads should receive pollen in order to ensure that all kernels on the cob develop. It is wind-pollinated. A few varieties have very high sugar content, hence the very sweet taste.

MAJOR PRODUCERS:

The major producers of sweet corn are U.S.A., China, Brazil, Romania and Yugoslavia.

STANDARDS:

There are no international standards for sweet corn. Many recommendations exist, e.g. U.S. Grade Standards, but they are not mandatory in international trade.

MINIMUM REQUIREMENTS:

Sweet corn should comprise intact, fully kernel-covered cobs with leaf wrappings which are intact, fresh and green and not spotted or damaged. The kernels should be yellow or golden, well formed all the way to the tip and sweet. The cobs should be free from infection by disease and insects and any foreign smell or taste.

Latin: Zea mays L. var. saccharata Koern.
French: Maïs doux
German: Mais
Spanish: Maiz

SWEET CORN

KEEPING QUALITIES:

Shelf life:

0°C, 90% RH, 4-8 days
20°C, 60% RH, 1-2 days

Ideal demands: 0°C (32°F), 90-95% RH

Recommended temperature:

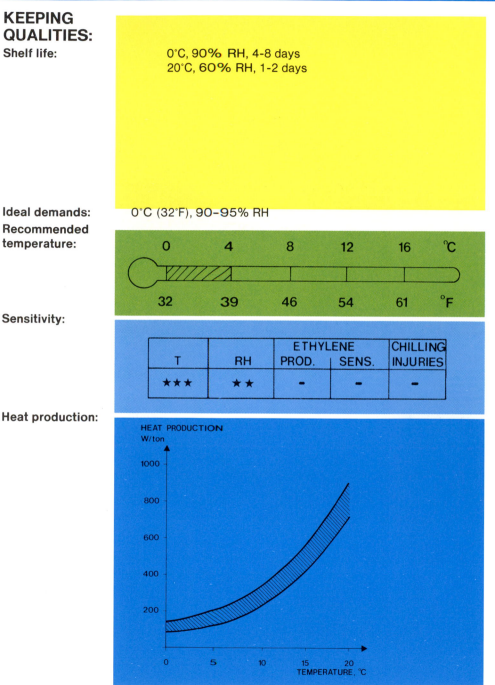

Sensitivity:

T	RH	ETHYLENE PROD.	SENS.	CHILLING INJURIES
★★★	★★	-	-	-

Heat production:

Specific heat: 3.30 kJ/kg x °C 0.79 Btu/lb x °F

Specific weight: Boxes, pallets 280-320 kg/m³. Bulk approx. 300 kg/m³

PEPPER, SWEET (Paprika)

DESCRIPTION OF PRODUCT:

Sweet pepper is native to Central and South America and brought to Europe by explorers, then on to Africa and Asia where similar types were already familiar. Sweet peppers are now cultivated in almost all tropical and sub tropical countries, but also in temperate countries in greenhouses.

The sweet pepper plant is a herbaceous annual. The fruit, which is the edible part, is bloated and hollow, and contains flat, yellow seeds. Sweet peppers contain the substance, capsaicin, which gives it its characteristic taste and which we know from cayenne pepper, paprika and chilli.

Sweet peppers exist in many shapes and colours of which the ripe fruit, the yellow and the red, are the mildest and sweetest.

MAJOR PRODUCERS:

The major producers are China, Spain, Rumania and Italy.

STANDARDS:

For trade within the EEC peppers should comply with EEC Standard No. 32. There are many other recommendations, e.g. U.S. Grade Standards, but they are not mandatory in international trade.

MINIMUM REQUIREMENTS:

Peppers should be intact, clean, fresh and sound. The fruits should be well developed, free from damage and blemishes and foreign smell or taste. They should bear no signs of attack by disease, insects, rot or mould, nor of chilling damage which may result in sunken parts often affected by rot. Peppers should be harvested with stalks.

KEEPING QUALITIES:
Shelf life:

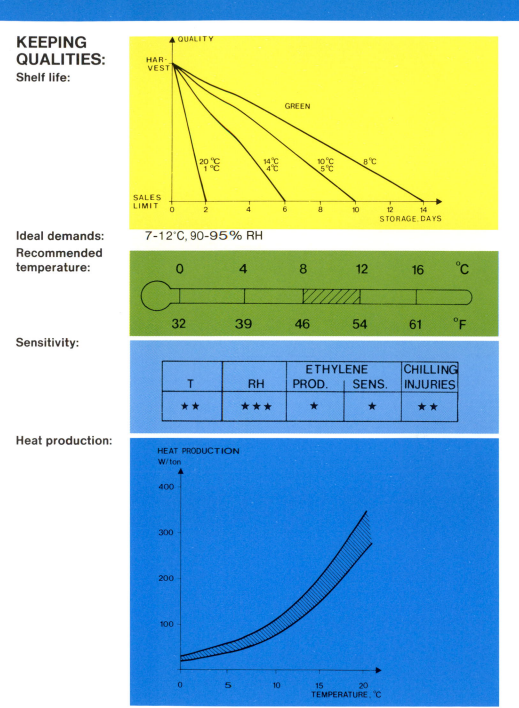

Ideal demands: 7-12°C, 90-95% RH

Recommended temperature:

Sensitivity:

	T	RH	ETHYLENE PROD.	SENS.	CHILLING INJURIES
	★★	★★★	★	★	★★

Heat production:

Specific heat: 3.98 kJ/kg x °C 0.95 Btu/lb x °F
Specific weight: Palletized boxes 180-220 kg/m³. Bulk approx. 280 kg/m³

POTATO SWEET

Latin:	Ipomoea batatas (L.) Poiret
French:	Patate douce
German:	Süsskartoffel
Spanish:	Batata

DESCRIPTION OF PRODUCT:

Like the potato, sweet potatoes probably originate from Peru, but in spite of their common name, they do not belong to the same family. From America it was spread to other continents, especially Asia.

The plant is a perennial vine, but is normally cultivated as an annual. It is not a short-day plant like the potato, but demands heat and a humid climate.

The root which is a swollen tuber can weigh up to 5 kg. It is irregular and oblong with a light brown or reddish skin. The flesh is pale or yellowish. Sweet potatoes contain approx. 5% sugar. They are very sensitive to rough handling.

MAJOR PRODUCERS:

The major producer of sweet potato is, by far, China, but Uganda, Indonesia, Vietnam and India are also important producers.

STANDARDS:

There are no international standards for sweet potatoes. There are many recommendations, e.g. U.S. Grade Standards, but they are not mandatory in international trade.

MINIMUM REQUIREMENTS:

Sweet potatoes should be intact, clean and sound. The surface should be smooth, without any signs of mechanical damage, bruises or cracks. Sweet potatoes should be free from signs of attack by disease, insects, mould or rot. The tubers should be firm and free from chilling injury, which can result in discolouration of the flesh and rotten spots. The flesh should have a colour, consistency and taste typical of the variety.

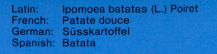

Latin: Ipomoea batatas (L.) Poiret
French: Patate douce
German: Süsskartoffel
Spanish: Batata

SWEET **POTATO**

KEEPING QUALITIES:

Shelf life:

14°C, 90% RH, 3-6 months
20°C, 60% RH, 2-3 weeks

Ideal demands: 12-16°C (54–61°F), 80-90% RH

Recommended temperature:

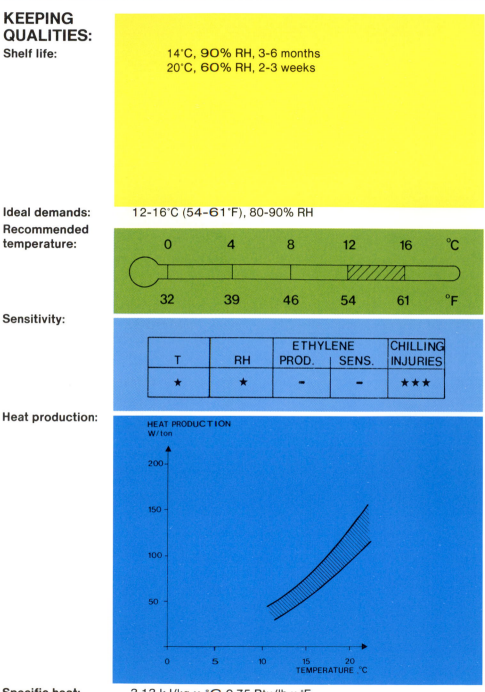

	°C
0 4 8 12 16	°C
32 39 46 54 61	°F

Sensitivity:

T	RH	ETHYLENE PROD.	SENS.	CHILLING INJURIES
★	★	–	–	★★★

Heat production:

HEAT PRODUCTION
W/ton

200 –
150 –
100 –
50 –

0   5   10   15   20
TEMPERATURE ,°C

Specific heat: 3.13 kJ/kg x °C 0.75 Btu/lb x °F
Specific weight: Boxes, pallets 300-350 kg/m³. Bulk approx. 600 kg/m³

233

TARO (Cocoyam, dasheen or colocasia)

Latin:	Colocasia esculenta (L.) Schott
French:	Taro
German:	Taro
Spanish:	Taro or Malanga

DESCRIPTION OF PRODUCT:

The wild forms of taro come from Burma and Assam where they may still be found growing in the swamps. In South East Asia it has been cultivated for several thousand years. From there it spread gradually throughout the entire tropical region.

The plant can be up to 2 m tall and is a perennial, but it is industrially cultivated as an annual crop. Propagation occurs by planting tubers or parts of tubers which have started to grow. From each of these planted tubers runners begin to grow, above or underground. These in turn develop new tubers.

Fresh taro are rather large tubers and they are covered with numerous ring-shaped leaf scars on the surface. The tubers contain a sticky milky sap. The inside of the tuber is white with either reddish or yellow spots.

MAJOR PRODUCERS:

The major producers of taro are Ghana, Nigeria, China, the Ivory Coast and Japan.

STANDARDS:

There are no international standards for taro.

MINIMUM REQUIREMENTS:

Taro should be intact, clean and sound. The tubers should be free from signs of attack by disease or pests. They should be firm, without roots and leaves, free from damage, spots or scars. There should be no discolouration of the flesh.

Latin:	Colocasia esculenta (L.) Schott
French:	Taro
German:	Taro
Spanish:	Taro or Malanga

(Cocoyam, dasheen or colocasia) **TARO**

KEEPING QUALITIES:

Shelf life:

12°C, 90% RH, 5 months
20°C, 60% RH, 2-4 weeks

Ideal demands: 11-13°C (52-55°F), 85-90% RH

Recommended temperature:

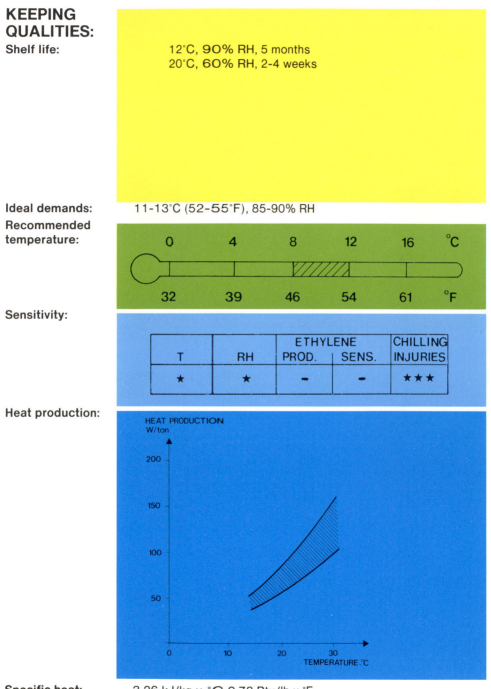

Sensitivity:

T	RH	ETHYLENE PROD.	SENS.	CHILLING INJURIES
★	★	-	-	★★★

Heat production:

Specific heat: 3.26 kJ/kg x °C 0.78 Btu/lb x °F

Specific weight: Palletized boxes 300-350 kg/m³. Bulk 550-600 kg/m³

235

TREE-TOMATO (Tamarillo)

Latin: Cyphomandra betacea (cav.) Sendtner
French: Tomate d'arbre
German: Tamarillo or Baumtomate
Spanish: Tamarillo or Tomate del arbol

DESCRIPTION OF PRODUCT:

Tree-tomatoes came from mountain areas of Peru. They have been cultivated in the tropical mountain regions of South America for a long time and later brought to Europe and South East Asia.

The plant is an almost 5 m tall tree. The leaves are big and heart-shaped. The flowers are pink and fragrant. The fruit is the size and shape of an egg. When ripe the skin is orange-yellow to red or brownish red. The ripe tree-tomato has yellow or red flesh. The ripe fruit yields to slight pressure at the stalk end.

The tree-tomato is divided into 2 chambers. The outer flesh which encloses the chambers is firm. The innermost part is very soft and contains small dark seeds. The juicy flesh is sweet-sour and tastes a little like tomatoes. The skin is bitter.

MAJOR PRODUCERS:

The main producing areas of tree-tomatoes are the northern part of South America and South East Asia.

STANDARDS:

There are no international standards for tree-tomatoes.

MINIMUM REQUIREMENTS:

Tree-tomatoes should be intact, clean and sound. They should be free from wounds, bruises, discolouration and mechanical damage and from signs of attack by disease and pests. They should also be free from rot which usually occurs at the stalk end and the stalks should, therefore, not be removed. The shell should not show signs of drying out. The flesh should be juicy and have a characteristic colour.

Latin: Cyphomandra betacea (cav.) Sendtner
French: Tomate d'arbre
German: Tamarillo or Baumtomate
Spanish: Tamarillo or Tomate del arbol

(Tamarillo) **TREE-TOMATO**

KEEPING QUALITIES:

Shelf life:

4°C, 90% RH, 3 weeks
20°C, 60% RH, 3-4 days

Ideal demands: 4°C (39°F), 85–90% RH

Recommended temperature:

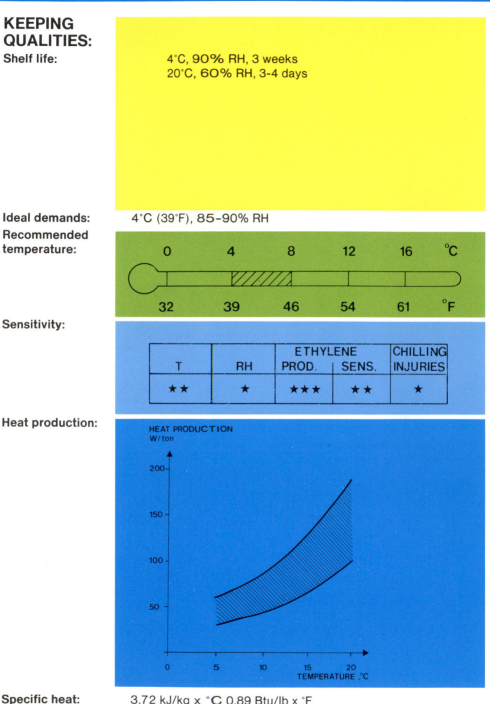

| 0 | 4 | 8 | 12 | 16 | °C |
| 32 | 39 | 46 | 54 | 61 | °F |

Sensitivity:

| | | ETHYLENE | | CHILLING |
T	RH	PROD.	SENS.	INJURIES
★★	★	★★★	★★	★

Heat production:

HEAT PRODUCTION
W/ton

Specific heat: 3.72 kJ/kg x °C 0.89 Btu/lb x °F
Specific weight: Boxes, pallets 300-350 kg/m³

TOMATO

Latin:	Lycopersicon lycopersicum (L.) Karsten ex Farw.
French:	Tomate
German:	Tomate
Spanish:	Tomate

DESCRIPTION OF PRODUCT:

Tomatoes originate from South America and have been cultivated in Mexico for a very long time.

Tomatoes are now grown all over the world - outdoors in temperate regions and in greenhouses in the colder regions. World trade in tomatoes and tomato products is enormous.

Size and shape depend largely on the variety. Diameters may range from approx. 3-10 cm. Tomatoes have a thin, tough outer skin. The inside is divided into several segments which contain the seeds.

MAJOR PRODUCERS:

The major producers of tomatoes are U.S.A., U.S.S.R., Italy, China and Turkey.

STANDARDS:

For trade within the EEC tomatoes must comply with EEC Standard No. 2. There are many other recommendations, e.g. U.S. Grade Standards, but they are not mandatory in international trade.

MINIMUM REQUIREMENTS:

Tomatoes should be intact, fresh and sound. They should be clean, free from foreign matter, smell or taste. Tomatoes should be firm, without spots, cracks, bruises or chilling injury which could result in a glassy appearance. They should be regular in shape and colour.

238

Latin: Lycopersicon lycopersicum (L.)
Karsten ex Farw.
French: Tomate
German: Tomate
Spanish: Tomate

TOMATO

KEEPING QUALITIES:

Shelf life:

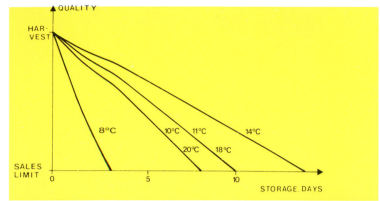

Ideal demands: 11-14°C (52-57°F), 80-85% RH For firm-ripe: 8-10°C (46-50°F), 80-85% RH

Recommended temperature:

0	4	8	12	16	°C
32	39	46	54	61	°F

Sensitivity:

T	RH	ETHYLENE PROD.	SENS.	CHILLING INJURIES
★★★	★	★★	★★	★★★

Heat production:

HEAT PRODUCTION W/ton

Specific heat: 4.00 kJ/kg x °C 0.96 Btu/lb x °F

Specific weight: Palletized boxes 330-400 kg/m³. Bulk approx. 560 kg/m³

239

DESCRIPTION OF PRODUCT:

The turnip was developed from wild cabbage, Brassica campestris, which is an ordinary weed native to the Old World. It was known in ancient Greece and, presumably, also in other parts of Europe.

The turnip matures quickly, in approx. 6 weeks, and is not particularly heat-dependant, which makes cultivation possible as far north as Greenland.

The turnip itself is flatly round, round, oval or elongated and the colour yellowish white. The turnip is normally white, but can be violet or green on the upper part. It should not be too large (max. 8-10 cm in diameter) as this will cause the root to soften and become spongy and possibly woody. Similarly, the taste varies depending on the variety and growing conditions. The turnip has a high Vitamin C content.

MAJOR PRODUCERS:

The major producers of turnips are Japan, China, Italy, France and U.K.

STANDARDS:

There are no international standards for turnips. Many recommendations exist, e.g. U.S. Grade Standards, but they are not mandatory in international trade.

MINIMUM REQUIREMENTS:

Turnips should be intact, clean and sound and the green top removed without damaging the turnip. Shape, size and colour should be typical of the variety and mis-shapen turnips should not be included. Turnips should be full-bodied and without any sign of drying out or of attack by disease, insects and rot. They should be free from foreign smell or taste. The flesh should be pale without being discoloured and woody.

KEEPING QUALITIES:

Shelf life:

0°C, 90% RH, 1-2 weeks
20°C, 60% RH, 2-3 days

Ideal demands:

0°C (32°F), 90-95% RH

Recommended temperature:

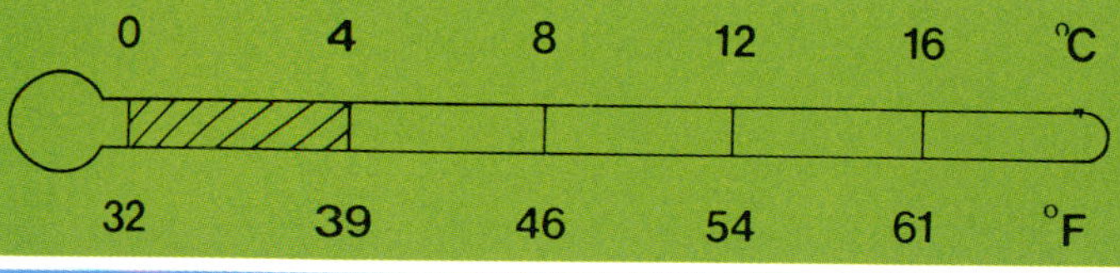

0	4	8	12	16	°C
32	39	46	54	61	°F

Sensitivity:

T	RH	ETHYLENE PROD.	SENS.	CHILLING INJURIES
★★	★★	-	-	-

Heat production:

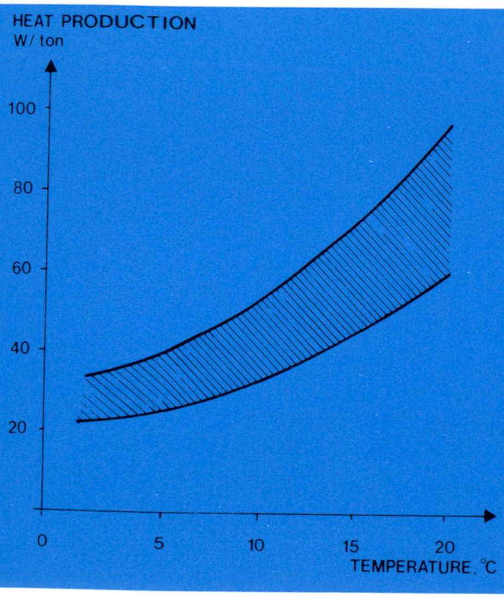

HEAT PRODUCTION
W/ton

TEMPERATURE, °C

Specific heat: 3.93 kJ/kg x °C 0.94 Btu/lb x °F

Specific weight: Boxes, pallets 280-350 kg/m³. Bulk approx. 250 kg/m³

241

Latin: Petroselinum crispum (Miller) Nym. ex A.W.
Hill ssp. tuberosum
French: Persil racine
German: Petersilienwurzel
Spanish: Chirivias

(Hamburg parsley or rooted turnip)
PARSLEY, TURNIP ROOTED

DESCRIPTION OF PRODUCT:

Turnip rooted parsley, closely related to parsley, is an ancient cultivated plant and originally native to the Mediterranean region, where it is still found growing wild.

Information on turnip rooted parsley is sparse and it is often mistaken for a parsnip, which it resembles in many ways. Turnip rooted parsley is generally smaller, less coarsely structured and has a finer and more subtle taste.

During growth turnip rooted parsley is very sensitive to poor soil conditions which may cause branching or uneven roots with ring formations. During mild winters turnip rooted parsley is able to withstand wintering in the fields, though it is more sensitive to frost than parsnips.

MAJOR PRODUCERS:

The major producers of turnip rooted parsley are West Germany and the East European countries.

STANDARDS:

There are no international standards for turnip rooted parsley.

MINIMUM REQUIREMENTS:

Turnip rooted parsley should be intact, sound, full-bodied and look fresh. It should also be clean, well-shaped without branching, free from attack by disease and pests, rust and other discolouration and from signs of mechanical damage or growth cracks. The roots should not be woody. If turnip rooted parsley is sold with leaves, they should be fresh and green.

Latin: Petroselinum crispum (Miller) **Nym.** ex A.W.
Hill ssp. tuberosum
French: Persil racine
German: Petersilienwurzel
Spanish: Chirivias

(Hamburg parsley or rooted turnip)
TURNIP ROOTED **PARSLEY**

KEEPING QUALITIES:

Shelf life:

0°C, 95% RH, 4-6 months
20°C, 60% RH, approx. 1 week

Ideal demands: 0°C (32°F), 90-95% RH

Recommended temperature:

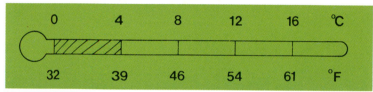

| 0 | 4 | 8 | 12 | 16 | °C |
| 32 | 39 | 46 | 54 | 61 | °F |

Sensitivity:

T	RH	ETHYLENE PROD.	SENS.	CHILLING INJURIES
★	★★	–	–	–

Heat production:

N.B. This diagram is for turnip rooted parsley with leaves.

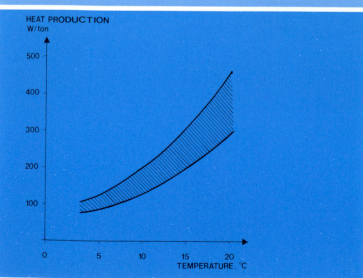

HEAT PRODUCTION
W/ton

TEMPERATURE, °C

Specific heat: 3.60 kJ/kg x °C 0.86 Btu/lb x °F

Specific weight: Boxes, pallets 280-350 kg/m³. Bulk approx. 250 kg/m³

(Japanese radish, mooli)
RADISH, WHITE

Latin: Raphanus sativus L. var. acanthiformis Makino
French: Radis japonais
German: Japanishe Rettich
Spanish: Rabano blanco

DESCRIPTION OF PRODUCT:

The white radish is considered to be one of the oldest cultivated vegetables. It is assumed to have its origin in the Mediterranean region and Asia Minor from where it spread to China and Japan. European, Chinese and Japanese varieties have been developed gradually. White radish derives from the Japanese type.

White radish can grow to 50 cm in length, half of which is underground. It is unnecessary to earth up around the roots to keep them white. This type of radish tends to shoot in summer as it is a long-day plant, but varieties that are day neutral are on the way.

The distinctive taste of white radish and other radishes is due to the presence of mustard oils. These compounds and the taste may vary considerably depending on variety and cultivation conditions.

The root is normally cylindrical with a tapering end. Its length is generally 30-50 cm and some roots may weigh up to 2 kg.

MAJOR PRODUCERS:

The major producers of white radish are China, Japan, West Germany, Italy and France.

STANDARDS:

There are no international standards for white radish.

MINIMUM REQUIREMENTS:

White radishes, should be intact, clean, full-bodied and appear fresh. They should be free from mechanical damage, deformities, branching, woodiness and disease and pest infection. If white radish is sold with leaves these should be fresh and green.

Latin:	Raphanus sativus L. var. acanthiformis Makino
French:	Radis japonais
German:	Japanishe Rettich
Spanish:	Rabano blanco

(Japanese radish, mooli)

WHITE **RADISH**

KEEPING QUALITIES:
Shelf life:

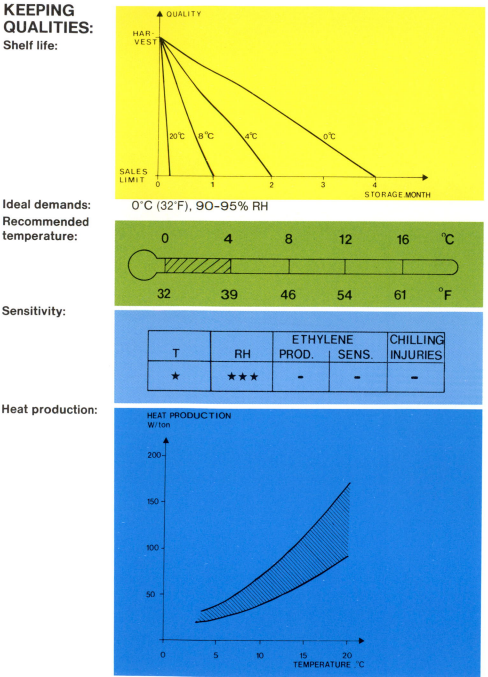

Ideal demands: 0°C (32°F), 90-95% RH

Recommended temperature:

Sensitivity:

	T	RH	ETHYLENE PROD.	SENS.	CHILLING INJURIES
	★	★★★	-	-	-

Heat production:

Specific heat: 4.06 kJ/kg x °C 0.97 Btu/lb x °F

Specific weight: Boxes, pallets 280-350 kg/m³. Bulk approx. 250 kg/m³

YAM (Dioscorea)

Latin: Dioscorea sp. L.
French: Igname
German: Yam or Igname
Spanish: Nana

DESCRIPTION OF PRODUCT:

In most parts of the tropical world yams grow freely out in the open and the various types of yam may be traced back to Asia, Africa or the Caribbean. Yam cultivation has been practised in Africa and Asia for over 5000 years and before that gathering wild yam tubers was quite common.

Yams play an important rôle as a basic nutrient in many tropical countries like manioc, sweet potato and taro.

It is a perennial, but is agriculturally cultivated as an annual crop. Small tubers or pieces of tubers are used for propagation. As the vine can grow 20-30 m it has to be supported with stakes. Yam roots are underground tubers that are formed on runners. Most varieties produce few, but very large tubers which often weigh 5-10 kg each. Imported yams are considerably smaller and come from varieties that produce small, but numerous tubers. The tubers are dark brown and very irregular. The bark is a horny layer with thin fibres. The flesh is pale and firm and has a neutral taste.

MAJOR PRODUCERS:

The major producers of yams are the Ivory Coast, Nigeria, Ghana and Cameroun.

STANDARDS:

There are no international standards for yams.

MINIMUM REQUIREMENTS:

Yams should be intact, clean and sound. The tubers should be free from any signs of attack by disease, pests, mould or rot and from soft spots, mechanical damage and bruises. Depending on type and variety, the surface should be smooth without deep depressions, cracks or wart-like growths. Yams should be firm, the flesh pale with no signs of rusting or other discolouration.

Latin:	Dioscorea sp. L.
French:	Igname
German:	Yam or Igname
Spanish:	Nana

(Dioscorea) **YAM**

KEEPING QUALITIES:

Shelf life:

16°C, 65% RH, 4 months

Ideal demands: 16°C (61°F), 60-70% RH

Recommended temperature:

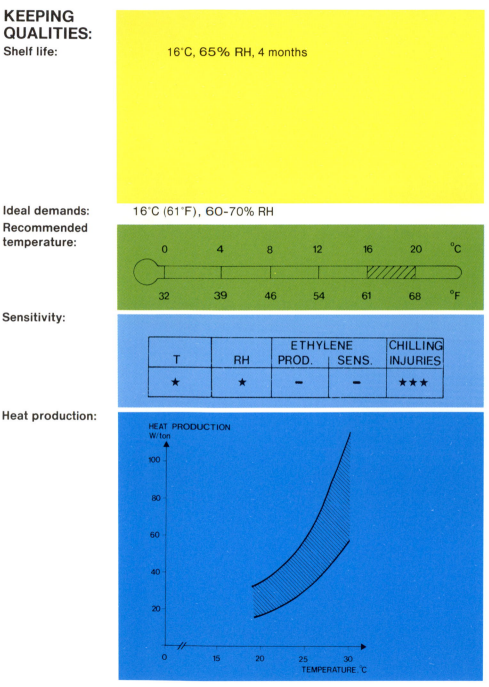

| | 0 | 4 | 8 | 12 | 16 | 20 | °C |
| | 32 | 39 | 46 | 54 | 61 | 68 | °F |

Sensitivity:

T	RH	ETHYLENE PROD.	SENS.	CHILLING INJURIES
★	★	–	–	★★★

Heat production:

HEAT PRODUCTION W/ton

Specific heat: 3.22 kJ/kg x °C 0.77 Btu/lb x °F

Specific weight: Boxes, pallets 300-350 kg/m³. Bulk approx. 600 kg/m³

247

INDEX

INDEX